WHAT WE LEAVE BEHIND

Also by Stanisław Łubieński in English translation

The Birds They Sang

Stanisław Łubieński

WHAT WE
LEAVE BEHIND

A Birdwatcher's Dispatches from the
Waste Catastrophe

*Translated from the Polish by
Zosia Krasodomska-Jones*

MACLEHOSE PRESS
QUERCUS · LONDON

First published in the Polish language as *Książka o śmieciach*
by Agora SA, Warsaw, in 2020
First published in Great Britain in 2022 by MacLehose Press

This paperback edition published in 2023 by

MacLehose Press
An imprint of Quercus Editions Ltd
Carmelite House
50 Victoria Embankment
London EC4Y 0DZ

An Hachette UK company

A CIP catalogue record for this book is available from the British Library.

ISBN (MMP) 978 1 52941 886 6
ISBN (Ebook) 978 1 52941 887 3

10 9 8 7 6 5 4

Typeset in Minion by Libanus Press, Marlborough
Printed and bound in Great Britain by Clays Ltd, Elcograf S.p.A.

Papers used by Quercus are from well-managed forests
and other responsible sources.

CONTENTS

INTRODUCTION

It was the first day of our holidays: July 2, 2018. The summer season was gradually warming up, but the beach still seemed empty and untouched. Everything looked perfect. Sand – unique Baltic sand, the best in the world – and the calm sea. But wait. Something was amiss. Something was wrong. The camera panning slowly over the beach stops and backtracks abruptly. It zooms in. A single sock lying on the sand. A white sock, scruffily rolled up as if on a teenager's bedroom floor. And with a dirty heel. That sock shattered at least two harmonies: the one reigning there on the beach, and the one within myself. I knew I couldn't just leave it where it was.

I had never been enraged by the sight of rubbish before. It was as though it lay, slightly blurred, in the blind spot of my field of vision. I only saw it out of the corner of my eye. I passed it indifferently – in the mountains, in forests, in marshlands. I wasn't the one littering, I didn't see why I should clean up after somebody else. But this felt like a line had been crossed: what I was seeing was plain incivility. In that moment I understood those people who, overcome by righteous anger and muttering expletives, tear down adverts from bus stops and street lamps. Somebody had spoiled this beautiful, sacred place – my favourite beach – with their disgusting dirty sock.

Or maybe it didn't happen so suddenly. Maybe this had been building up inside me for years? Maybe my anger at all those damned plastic bags we keep in endless bundles at home simply boiled over? My anger at the throwaway junk from "Everything for 0p" shops, the cheap, useless garbage sold at seaside kiosks. At the heaps of rubbish on roadsides or illegal dumps in forests we're so accustomed to seeing they don't even seem out of place.

But it was the sock that made my blood boil.

I didn't spend the rest of that afternoon craning my neck to glimpse the sandwich terns that sometimes fly overhead. I set off, my eyes glued to the sand. I didn't have a bag, so I stowed my grim harvest inside the sock. The first item was a braided fishing line. At the time, I knew very little about the tangles of nets that float around the world and catch whatever drifts past in their mesh. Next were a few balloons, partly buried, with just a faded ribbon poking out of the sand. It didn't occur to me that they resemble squid, which is why many seabirds eat them. I didn't realise they can easily cover up to a thousand kilometres on the wind. Over the next few weeks, I found dozens of balloons.

There was also an inoffensive plastic ball from a toy gun. There were sweet wrappers, a crisp packet, some cling film – in other words, food packaging, common beach waste. And a plastic fork, which only vaguely brought to mind the issue of the billions of disposable single-use products we use for a minute or quarter of an hour. I found bits of beach windbreakers, a wall plug, more fishing lines and a strand of material from one of those construction bags that disintegrate into thousands of threads. It was the first time I had studied every centimetre of the beach so closely, and I realised that mingling among the infinite grains of sand, there were equally uncountable specks of waste. I knew my spontaneous clean-up was pointless, that the rubbish was ubiquitous, but I couldn't stop.

I dug up a plastic glove for prodding bread in shops, probably used only once. I found a wrapper from some pocket tissues. "Hygienic," it read. "Thorough cleansing" – that was on a label for face wash. There was a pack of wet wipes, Moist Towelettes, that had come all the way from Florida. That sunbather didn't like having grimy hands. It struck me how many of those products said, "We want cleanliness, we don't want dirt or bacteria". As if we were trying to cut ourselves off from our surroundings with a sterile barrier. But the litter stayed behind long after we'd gone home. We wanted to be clean and free of germs; somebody else would clear up the leftovers, wouldn't they? Or the sea would take them.

I managed to cram twenty-eight objects inside the sock. And each had its own story to tell.

In that short time, my eyes were opened to a whole new galaxy of topics. I felt slightly helpless – tidying up wasn't enough. But one's own powerlessness is hard to bear. I wanted to do something. Something within my capabilities. I felt I had to write about it. My "Litter of the Month" column in the online journal *Dwutygodnik* began with a piece about two of my finds: the sock and a bag from a Bulgarian duty-free shop I picked up the next day. After that I started seriously collecting rubbish and keeping the items I found most interesting. I litter-picked on beaches, but also in the Sudeten mountains, the Roztocze hills in central Poland and the Suwałki region in the northeast. And on it went.

On a bird-watching trip, we admired the western marsh harriers and struggled to free binder twine pressed deep into the surface of a dirt road. I brought back a licence plate from a visit to some ponds that are home to common goldeneye ducks. In Miedzianka we found an old basin, some bottles and a heap of mouldy sandwiches. We didn't take those. They were surely the traces of the average hunter's legendary concern for wild animals. Fortunately, apart from the sandwiches, we also saw some crossbills and a honey buzzard on our walk.

Over time, I discovered that, rather like birds, waste has its own favourite habitats; forest rubbish is often different from that found by rivers, on beaches or in meadows. I wasn't naïve – I knew it hadn't ended up there by itself.

Now and then, I came across exotic and unexpected rubbish. I noticed that the sea, following some algorithm known only to itself, sends forth Russian or Swedish or sometimes toxic waste – like a large tub of petroleum-based solvent. I read the sand-worn labels: "irritating to skin, respiratory system and eyes"; "may cause long-lasting harmful effects to aquatic environments"; "protective equipment must be worn when handling". How did a bucket of this poison end up in the sea?

And socks? I discovered that they crop up quite regularly on beaches. But not as regularly as the objects I would find on a daily basis. Familiar items that I'd be known to use myself entirely unthinkingly, never wondering what lay in store for them next. Single-use cotton buds, for instance. The cotton vanished, leaving behind a blue, transparent tube rolling about on the strand line. What happens to a little stick like that when you throw it away? How was I supposed to know? I'd just hoped that someone out there would do something sensible with it.

I thought it was worth taking a closer look at all this. To take a look at myself. My rubbish. The things I would buy and throw away. I began to collect not just rubbish, but other intriguing objects I came across on my walks. Hipster teabags made of an unknown material, an earthenware owl found in a pile of burned cables, a box for pasta with a little transparent window, the fan from a Seat Ibiza, a biodegradable shopping bag from Italy, a coil of cords collected from the beach. I didn't, however, take even one of the hundreds of funeral wreaths I found on the edge of a field outside Warsaw.

I acquired the habit of rummaging in my apartment block's bins, patiently transferring dirty polystyrene trays and even snotty handkerchiefs to the black container for general waste. I started measuring and counting my own rubbish, without really knowing where it would lead. I looked at packaging. I wanted to learn something about the materials it was made from and what might happen to it next. I started reading about recycling, the history of consumerism, I spent hours poring over industry reports. I contacted experts. They told me about flint axes and which bottles are deadliest for forest beetles. They let me watch optical sorters in treatment facilities and study polymer ravines through a microscope. It felt like I was seeing the far side of the moon for the first time, a parallel world running entirely beyond my, or our, notice. A book was the only possible outcome.

With each day, the price of our ignorance, our insouciance, our thoughtlessness grows. Just as the natural odour of our bodies is masked by deodorants and our waste is flushed away down sewage pipes, our rubbish vanishes, carried off by dedicated services. We don't give it much thought, we put it out of our minds. But what we're creating isn't mere untidiness, a mess we can clean up. Change won't come from determined individuals, keen altruists who pick up three bottles on their morning jog. Earth Days won't help, even if they were to happen every day. Our world is overrun by waste, which reaches virgin beaches and the most isolated corners of the planet. Change will only come from a profound shift in our awareness and our habits, and finally, bold legislation to force those who bury us in a daily avalanche of rubbish to face the consequences. There's a lot of work ahead for our lazy, lumbering, convenience-driven civilisation. We can't put it off any longer. Let's get to work.

JULY 3, 2018

c.4.00 p.m., 54°49'55.4"N, 18°08'25.4"E

Wherever there's rubbish, there'll always be a plastic bag. Just take a good look around. But this one was special: not some ordinary, everyday bag, but a sturdy yellow specimen bearing the exotic words, "Duty Free Store Varna & Burgas". Bits of waste can lead long and interesting lives, and many are well travelled. I found myself daydreaming about my Burgas bag's numerous adventures on its journey from the Black Sea to the Baltic. What had its owner been up to on the Bulgarian riviera?

With regret, I've had to abandon my fantasy that the Burgas bag had circumnavigated Europe. Thick polyethylene bags like this one will sink, sometimes to quite remarkable depths. As I later learned, in April 2019, Dallas businessman Victor Vescovo descended about eleven kilometres deep in a bathyscaphe submersible, and at the bottom of the Mariana Trench he spotted what looked like a carrier bag.[1]

The Burgas bag became a key element of my litter-picking arsenal. I filled it repeatedly until, in the end, it fell apart, riddled with holes from plastic straws. It disintegrated but didn't disappear – that's what's most precious and most terrifying

about plastic bags: they're indestructible. A polyethylene bag is sure to outlive us. In a way, it's perfect: both light and extremely durable. Thirty years ago, the Burgas bag would have been treasure. Its owner would have washed it, ironed it, smoothed it contentedly, taken it to birthday parties. Today it's worthless. Carrier bags are usually discarded after a single outing. It's estimated that on average they're used only for around a quarter of an hour. According to data cited by the United Nations, between one and five trillion polyethylene bags are consumed every year.[2] The scale of production – and the resulting scale of pollution – is vast.

Carrier bags cause such widespread concern that they've even been banned by Al-Shabaab, a cell of Al-Qaeda. They are prohibited in many countries or subject to charges. I also try to avoid them, but it only takes a moment of inattention and I have another to add to my collection, from the pet shop, the bakery, or the vegetable stall. It feels stupid to throw it away so I stuff it inside another bag. Right now, we have fifty carrier bags in various cupboards of our kitchen.

Invented in the 1950s, plastic bags really took off in the middle of the following decade. That was when the Swedish firm Celloplast obtained a patent for the American market. In the 1980s, the major American supermarket chains started using them and they quickly supplanted the paper and re-useable bags of the past. Forty years later, in an era of growing environmental awareness among consumers, some chains have started abandoning them. The British supermarket Morrisons decided to replace plastic bags with paper ones in their fruit and vegetable aisles.[3] For a time, Tesco stuck with plastic because it has a smaller carbon footprint.[4] The debate over which is better, plastic or paper, is basically pointless. You don't treat the plague with cholera. The best bag is one that doesn't exist.

1

NINETEEN CARTRIDGE CASES

The noise of Provincial Road 177 fades away. It's a quiet single carriageway, but in the silence of the late-summer pine forest, the presence of every passing car is long felt. The russet tree trunks stand in ruler-straight lines. Orderliness reigns over the tree plantation. No shrubs, no glades, no crooked hornbeam. The most frequently sighted animal here is man, *homo sapiens*, in pursuit of *macromycetes* from the *fungi* kingdom, more commonly known as mushrooms. But not today, and nor tomorrow, because it hasn't rained here for months. Twigs crunch underfoot, as do an old moss-covered umbrella and a rusty number plate. In the distance, where the sun is slowly setting, the forest thins. Through the trees I see sky, wind and space.

There's a twenty-metre slope and a large pond encircled by reeds. Or rather, several ponds split by a causeway, which somewhat disrupts the well-ordered monoculture. I follow animal tracks down to an alder copse. There are traces of sharp little hooves in the mud – deer come here to drink. A goldeneye watches me suspiciously from the water. That eponymous eye stares unblinkingly: the duck is frozen, waiting for my next move. The alder trunks cast dark shadows over the water. Standing on the overgrown path, I bend to pick up a red object in the grass. An oval tube that looks like it's made from low-density polyethylene (LDPE), with a metal base. It's fresh and uncorroded. A cartridge for shooting ducks: 7-cm case, 3-mm pellets, 32-gram load. Manufacturer: Pionki Hunting Ammunition Factory. The goldeneye flies away.

*

The bird hunting season begins in mid-August. Experts say that's too early, females can still be seen tending their young in September. Four duck species may be shot, and although the goldeneye is not on that list, its caution is warranted. Hunting

not only means death and injury, it also brings stress, forced migration, interrupted feeding and the reduced immunity and reproductive capabilities that this entails.[5] Every year, as the season begins, a cannonade erupts at ponds and lakes across Poland. It is permitted to kill the good-natured mallard that we know from parks, the timid Eurasian teal, the tufted duck with that fanciful crest on the back of its head, and the silvery pochard.

Let's start with a clarification to avoid any misunderstandings: hunting ducks has no practical justification – it's purely for sport. A display of dexterity, like shooting live clay pigeons. But what have the ducks done to deserve to die? Pond owners sometimes complain that they eat fish food. They certainly do. According to studies from the 1980s, they eat between 2 and 7.5 per cent of distributed feed. That's not very much. Scientists say that the presence of many bird species at ponds brings advantages that outweigh the drawbacks. Ducks, coots, grebes and even herons prevent the surface water from becoming overgrown, they eat the larvae of predacious insects that feed on spawn and fry, and they clear sick or dead fish from the surface.[6] Why do we kill ducks, then? Simple: it's tradition.

And tradition – much like the fatherland – cuts short all discussion and clamours instead for war. This is what our forefathers did, don't you dare question our forefathers, they couldn't have been wrong. What does it matter that the world has moved on?

It's no secret that most hunters know little about birds. Identifying flying ducks is a difficult task, particularly if you don't have a pair of binoculars and a wealth of experience around your neck. Who'd want to patiently study bird atlases and then spend years laboriously acquiring knowledge on the ground? On a hunt you only have a few seconds to take your shot, a moment later the bird is too far away for the pellets to pierce its skin. On hunting forums and even in official reports of the Polish Hunting Association (PHA), you often come across the term "wild ducks", usually without any mention of the actual species. A "wild duck" could be any duck. "Wild duck" ignores the fact that some of them are rare and endangered. "Wild duck" makes a mockery of conservation regulations.

Why would a hunter bother to learn how to spot a female of the protected and extremely rare ferruginous pochard species when it isn't required for the exam?[7] All that's checked is theoretical knowledge. I did the test on the PHA's website, and most of the hundred or so questions were about customs, jargon, dogs, organisational structures and the basics of using weapons. There was one question about ducks, but nothing on the identification of species. It was about game birds – as if the others didn't matter, as if the pellets had no effect on them. No-one checks your ability to identify birds in practice. That's probably why protected species, sometimes rare and endangered ones, are killed alongside game species. Not everybody wants to eat hunted ducks either – not everybody revels in the prospect of lead pâté. Many corpses and wounded birds that can no longer fly are left on the water or in the reeds, even though hunters have an obligation to find them, kill them if necessary and take them away.

In 2019, the Polish National Committee of the International Union for Conservation of Nature (IUCN), which brings together governmental and non-governmental environmental organisations from one hundred and sixty countries, called on the Polish authorities to remove three duck and coot species from the list of approved game birds. Detailed analysis suggests that their populations in Poland are at risk of collapse. Some are near the point of no return, and not just in our country. The pochard, for instance, is a threatened species at global level. Only three other breeding birds with that status appear in Poland: the extremely rare greater spotted eagle, the European turtle dove, which is disappearing at an alarming rate, and the aquatic warbler, for which ornithologists travel to the Biebrza River from all over the world. All of them, except for the pochard, are protected species. Meanwhile, the number of hunted tufted ducks has declined by 75 per cent in the last three decades, while hunting the Eurasian teal may endanger the very similar but rare and protected garganey, leaving it in the same situation as the Eurasian coot, a bird from the rails family, whose Polish population has dropped by 90 per cent since the 1980s.

If the Environment Ministry , the Chief Nature Conservator and the Polish Hunting Association, to whom the appeal was addressed, had agreed with the experts' conclusion, only the mallard would remain on the list of game ducks. The IUCN stresses the need to shift the hunting season so that shooting does not begin when adult birds are still tending their flightless young. This prolongation of the breeding season is the result of climate change and is a widespread and advancing phenomenon. The IUCN also recommends a ban on hunting after dusk when protected and game species are indistinguishable even for experienced ornithologists. The Union further calls for detailed registration of killed birds, and thus for an end to the imprecise concept of "wild ducks" and "wild geese", as well as to the use of lead hunting ammunition on account of its toxic properties.[8]

Perhaps the Union should also call for cleaning up after oneself? Hunters enthusiastically claim that hunting puts them in touch with nature. So how are they not offended by the sight of red and green fragments of plastic, the same mess their brethren left at the ponds by Provincial Road 177?

*

The Zatorskie Ponds near Oświęcim are famous for their fish farms, but they are also one of the most valuable bird hotspots in southern Poland. For this reason, the ponds and their surroundings, known as the Lower Skawa Valley Natura 2000 area, are protected. Rare duck species raise their young amid the extensive reeds; the overgrown islands are home to Poland's largest breeding colonies of the small, nocturnal black-crowned night heron. But the Zatorskie Ponds, a true bird paradise, is also part of Hunting Zone No. 99. In recent years, around two thousand "wild ducks" and coots have died here annually.[9] If we take a broader perspective, the Natura 2000 area is split between Hunting Zones Nos. 79, 100, 117, 123, 124 and to a lesser extent also 78, 80 and 116. A further one thousand eight hundred coots and "wild ducks" die in these zones every year. So, while most shooting takes place at the Zatorskie Ponds, ducks who flee in terror from the booming gunfire risk flying into the crosshairs of hunters in the surrounding area.

Between 2008 and 2014, studies were carried out at the Zatorskie Ponds into the causes of bird mortality. Two hundred and sixty shot birds were found in the reeds, on the water and on causeways.[10] Almost 40 per cent belonged to highly protected species. Many of them bore no close resemblance to game ducks, coots or geese. The most frequently killed bird was the black-headed gull, but white-tailed eagle and black-crowned night heron were also found, as were swans. Relatively common birds and extremely rare ones died – in total, as many as eighteen protected species. Birds were killed all year round, although most fell victim during the hunting season. Who was doing the shooting? Poorly trained hunters or skilled poachers? Did their fingers tremble with excitement as they pulled the trigger, or were they calm and sure? Did they shoot for the adrenaline high or out of a sense of duty, a sincere belief that all pests and weeds, anything not subject to man's control, must perish?

In Poland and many other European countries, anti-hunting movements are growing. During autumn hunts in the 2018 season, ornithologists observed hunters for the first time as part of a civic monitoring campaign. They managed to capture on camera the moment a protected garganey was shot down, but they weren't allowed access to the hunting grounds, and by the time the police arrived the hunters had probably discarded the slaughtered bird, leaving them with nothing to do but seize the memory card containing the records of the event. By the end of the season, the birdwatchers had spotted eighteen discarded corpses in the area.

It's very difficult to catch someone red-handed, as the fishponds where hunts take place are usually closed to outsiders. Witnesses say that in some locations, shots are fired at anything that flies. Social media provides some fragmentary evidence of the problem. In 2018, a photo of two smiling women posing with shot ducks in west Poland sparked controversy. One of them was holding a dead garganey, a protected species. The case was dismissed – apparently it was impossible to establish who actually killed the bird. Pellets can't be attributed to a specific gun, so participants in group hunts can take aim with impunity.

In August 2019, a hunter from the province of West Pomerania posed for a photo with a shot gadwall. Once his friends had pointed this out, he deleted it from his profile. Clearly, he was unable to identify a protected species even when holding it in his hand. Following the appearance of compromising photographs and recordings (such as the abuse of a dying deer), the PHA called on its members to avoid publishing material from hunts on the internet, but in any case, hunters are more likely to brag about their achievements in closed groups and trusted circles.

*

In late summer and in autumn, the causeways between ponds are thick with cartridge cases. Many hunters don't clean up after their visits, considering themselves guardians of a very different order. Guardians of tradition – and people with such a noble calling don't need to bend down to pick up litter. A few steps from the red cartridge lies a green one. Italian Cheddite ammunition, 34 grams, 3.5 mm pellets. The metal is a bit rusty – it's probably been there since last year. Who shot it? One of the good people of W. county town? Perhaps Mr. D., owner of a wholesale outlet? Or perhaps one of the foreign guests of Mrs. S., who organises the shoots? Or maybe the owner of company T., who bought his son H. an airgun for his birthday? Let the boy learn how to be a man. Did the hunter creep here at dawn or did he shoot in the evening, barely able to spot the birds as they took flight?

One thing is certain. A finger pulled the trigger, the firing pin hit the primer, the gunpowder caught fire, and the gas sent over one hundred and thirty pellets down the barrel. That's how many are crammed into one of these Cheddite cartridges. At 35 metres, the pellets cover an area around 2.5 metres in diameter. Thirty-five metres is the maximum allowed distance to shoot at a target. But who's so pernickety in real life, when the adrenaline is coursing through one's veins? At 50 metres, the dispersion range is 4 metres. If a hunter shoots ducks as they take off from a pond or lake, he can hit several birds at once. One or maybe two will fall into the water, the rest will fly away, carrying bits of lead in their bodies. Dozens of little balls will sink to the bottom or fall on the bank or causeway. To kill one bird, a hunter needs on average between six and ten cartridges.[11]

Most of the time, the pellets remain in the breast and wings. Even a small amount is a death sentence. The bird will be unable to escape predators or it will simply be poisoned by the lead. But they don't have to be shot to suffer lead poisoning. Waterfowl swallow pellets they find on the shore or at the bottom of lakes and ponds – they mistake them for small pebbles, known as gastroliths, that grind food in the gizzard like millstones. A study showed that, in the south of the continent, the highest number of pellets were eaten by pintails. They belong to a group known as dabbling ducks, which means they upend on the surface of the water to feed. Birds like pintails are particularly at risk in muddy, shallow waters, where the pellets remain within their reach. But swallowing lead is also a problem for diving ducks, since they look for food at greater depths and will pick up any pellets that have sunk to the bottom. Studies in Europe have shown that on average goldeneyes and tufted ducks have the highest quantities of lead in their guts – each accounting for around 12 per cent of birds examined. Record levels were found in tufted ducks – pellets were found inside 80 per cent of birds at one site in Spain and in 58 per cent in Finland.[12]

Some species are poisoned by lead "inadvertently", so to

speak. Predatory birds and scavengers swallow pellets lodged in carrion or the meat of their prey. The first article on this topic was published in the late nineteenth century, while according to the most recent findings, pellets are found in the stomachs of over one hundred and thirty bird species.[13] Lead ammunition in carrion was the main cause of death for the critically endangered California condor, which was pushed to the brink of extinction in the late 20th century. In 2008, hunting with lead pellets was banned in the condors' distribution area,[14] and in 2019 the population of the species, in the wild and in captivity, stood at around five hundred birds.

*

Hunters often say that shooting ducks is a way for them to get healthy, "chemical-free" meat, proudly declaring that they feed their families with it. But studies of lead and cadmium concentrations in the bodies of mallards and Eurasian coots shot at the Zatorskie Ponds have provided an interesting counterpoint. In one third of birds analysed, the levels of both elements in their blood exceeded European Union limits for human consumption.[15] The birds were poisoned by long-term contact with pellets. It's hard to say if they were exposed at the ponds or if they consumed lead elsewhere before heading there during the autumn migration season. The study demonstrates that the poisoning of waterfowl with heavy metals is widespread, and while it's estimated that there are roughly three pellets per square metre at the Zatorskie Ponds, that's nothing compared to the extreme concentrations found for instance in Spain. At one site on the Ebro delta, almost two hundred and seventy pellets were counted per square metre. Mortality as a result of lead poisoning among mallards that wintered there was 37 per cent.[16]

*

It's not only lead that poses a problem, but the very practice of bird hunting. In Poland in 2019, 183,000 birds were killed in compliance with the law.[17] But these figures don't cover the hundreds of thousands of injured birds that fell into the reeds and were left there by hunters. There's no way of counting how many die from stress or are flushed out of bushes and killed by predators, or how many orphaned young don't survive without their parents. The Polish Hunting Association asserts that hunting affects barely a thousandth of the population.[18] In its report, it states that hunters care for game species and that their numbers recover each year.

It's hard to say what indicators the Association is using for its claims. Hunters make no inventories and no local assessments of duck and goose populations, which are always presented without distinction between species. They don't count coots or woodcocks. The Polish Hunting Association cannot state that the game population of the greater white-fronted goose recovers, because the bird's nearest breeding area lies in the sparsely populated, remote forests of the polar tundra in Nenets Autonomous Okrug. Annual hunting plans are drawn up approximatively. The well-balanced idyll presented in the reports has nothing in common with reality.

For years, the impact of hunting on the environment was

underestimated. Other ecological problems seemed more alarming: environmental transformation, destruction of habitats, the use of chemicals in agriculture. A few years ago, the organisation BirdLife International published a shocking report showing that poachers kill up to thirty-eight million birds every year across Europe, the Middle East and North Africa.[19] It turns out this is just the tip of the iceberg. In 2018, the Committee Against Bird Slaughter (CABS) took a closer look at the data on legal hunting. They found that in EU countries (plus Norway and Switzerland) alone, in full compliance with the law and established limits, nearly fifty-three million birds are killed every year.[20]

Legal hunting goes largely under the radar of public opinion and nobody dares to question its legitimacy. Across the EU, a total of eighty-two bird species may be hunted. National regulations vary in this respect: in Poland thirteen species may be shot while in France that number is sixty-seven. Here's one striking example of the harmful impact of hunting on bird populations: every year, one and a half million European turtle doves are killed, a species that the International Union for Conservation of Nature says faces a high risk of extinction in the wild. Over the last thirty years, the global population of European turtle doves has declined by 80 per cent.

A grave problem is the habitual infringement by certain Member States of Article 7, paragraph 1 of the Birds Directive. That provision warns that hunting must not thwart efforts undertaken to protect a species in its distribution area.[21] These rules are inconsistently applied, however; EU money is spent on protecting the Eurasian curlew in Poland, but the bird can legally be shot in France. By way of example, in the 2013–2014 winter season, seven thousand curlews were shot on the French coast. That's the official data, but of course many more may have fallen victim unnoticed.

In Poland, the population of these birds is declining, now estimated at two hundred and fifty pairs. And our curlews die in France, too. In late August 2019, the GPS tracker of a young curlew with leg-ring J64 stopped emitting a signal somewhere near Saint-Valery-sur-Somme. This area is equipped with highly developed hunting infrastructure, with hides and ditches dug to lure birds that feed near the sea. There is no doubt that the curlew was shot. It met with the same fate as other birds whose trackers fell silent in previous years. And only a small part of the Polish population is monitored in this expensive way. We don't know how many curlews fitted with simple plastic leg-rings die in France every year.

I doubt French hunters realise how much trouble it takes to rear a single Polish curlew. Every year in early spring, ornithologists identify nests and enclose them in electric fencing. They ask farmers not to mow their meadows too soon. When eggs are laid, the ornithologists take them and place them in incubators. The real eggs are replaced with speckled, olive-green painted wooden dummies. If predators damage the fake eggs and the adult birds abandon the nest, the chicks are raised in aviaries and released once they are able to fly. Curlew J64, lost over

France, was one of those reared in an aviary. He was released in the middle of July 2019, not far from his family nest in the Upper Biebrza valley. If the dummies aren't disturbed, they are swapped for the real eggs just before hatching, so the chicks can be born in the wild. And then come six long weeks of crossing your fingers that the young, still flightless bird isn't eaten by a fox. If it survives, it will spend its first winter on the coasts of Western Europe. All that's left to do is pray it doesn't pick the French coast.

*

I'm staring at the ground, so I'm not looking where I'm going. I keep walking into huge spider webs that stretch across the path. It seems no-one has come this way for several days. The feathers of a grey heron and a few bits of wing are scattered by the edge of the water. A fox must have got to the corpse, but who actually killed the bird? A dozen or so cartridges lie on the causeway between the ponds. They stand out because there isn't really any other rubbish here. They're marked FAM Pionki (Hunting Ammunition Factory); 28, 30, 32, 34 grams. Pellets in the reeds, in tree trunks or at the bottom of ponds will take up to three hundred years to disintegrate. Every year, hunters in the EU leave up to fifty thousand tonnes of lead in fields, forests and bodies of water. In the United States and Europe, hunting ammunition is considered the most significant unregulated source through which this toxic element is released into the environment.[22][23]

In the silence I hear the sound of a grebe's maniacal cackle coming from the pond. Who knows, perhaps it too carries pieces of lead under its skin? Patient, relentless and lethal. Lead poisoning is not a new phenomenon – it was known to people in Antiquity. In the second century BC, Nicander,[24] a Greek poet and doctor, described the effects of white lead: dry throat, paralysed tongue, shivers and colic (all symptoms of lead poisoning, also known as saturnism). Caesar's architect and designer of siege engines, Vitruvius, warned that lead piping in aqueducts was harmful to public health.

In the 1980s, a rather daring hypothesis was posited, asserting that the fall of the Roman Empire was due to widespread lead poisoning.[25] Defrutum, a grape syrup used to sweeten wine, was cooked in lead pots and the element was also used in cosmetics and medicines. According to Professor Jerome Nriagu, the Roman elite were particularly exposed to its effects. For many centuries, the substance impaired their reproductive capabilities and their intelligence, and was probably behind much of their strange behaviour. We've all read about the legendary perversion and pathological cruelty of Roman emperors. Were Caligula and Heliogabalus the victims of lead poisoning? Did Rome go soft in the head until it was conquered by barbarians? Professor Nriagu's theory is nowadays treated more as a curiosity, but saturnism certainly was rife throughout ancient Rome and it is known to be heritable. Raised levels of the element have been found in the blood of newborn Inuits whose parents had eaten hunted birds.[26]

*

The World Health Organisation's website says that lead attacks the central nervous system but also the liver and kidneys. It accumulates in bones and teeth, and during pregnancy it is released into the blood stream and influences the development of the embryo. It is especially dangerous for children, since the young body absorbs four to five times more of the substance than adults do. Lead poisoning in children impairs brain development and increases antisocial behaviour. These changes are often irreversible. In many countries, the predominant source of poisoning is the inhalation of fumes from smelting and improper recycling of lead. The element also enters the body through the digestive system – together with polluted soil and dust, through food containers and water supplied by lead pipes. The WHO considers it to be one of the ten most dangerous chemical substances. There is no level of lead in the blood that is considered safe. In 2016, the element caused the death of half a million people worldwide.[27]

*

The sun sets behind the trees. In the silence, a great egret flies slowly past like a ghost against the darkening forest. The ducks tuck their beaks beneath their wings and fall into a vigilant sleep. Centuries pass, the lead is still going strong. Pb, number 82 on the periodic table. Since I left secondary school, new exotic elements have been discovered, such as Copernicium, Moscovium and Tennessine. Leaded petrol and lead white paint have disappeared from our country but just like hunting, lead ammunition persists, even though there are equivalent, cheap, nontoxic substitutes, such as steel pellets.[28] At those small, little-frequented ponds I found nineteen cartridge cases with pellets of different diameters and weights. Nineteen – it may not seem like a lot, but they contained around three and a half thousand lead balls. What is hunting with lead pellets if not the authorised pollution of the environment with a neurotoxin? That might sound like a provocation, but it's the honest truth.

AUGUST 31, 2018

c. 2.00 p.m., 54°49'57.2"N, 18°08'49.8"E

I have a weakness for rubbish with Cyrillic lettering, so I was obviously delighted to spot a small black tube on the sand with the label "Penilux" – Интимная гель смазка ["intimate lubricant gel"] and a symbol representing a penis. It's a gel that "guarantees comfort during intimate encounters". Typical multi-layer packaging, impossible to recycle. This wasn't something the sea had picked out for me. Penilux hadn't been carried here by ocean currents from the Curonian Spit or the Gulf of Bothnia. I found it lying by a sand dune – somebody out here in this secluded spot, far from the nearest resort, had enjoyed an active holiday. I read online that, in one month, Penilux increases the length of the penis by five centimetres and enhances sexual experiences, thanks to the beneficial extracts of horse chestnut, Japanese quince, ginseng and some kind of Amazonian shrub called Ptychopetalum.

I figured it must be effective since it costs more than the limited-edition "Rasputin" cream. Ultimately it was Alexei, age 34, who convinced me: "When I discovered that you could increase the size of your penis, I ordered the gel immediately! I

applied it for one month, and though admittedly I sometimes missed a dose, the result was an extra 2.5 centimetres and a cast-iron erection". I decided to find out where this revolutionary medicine was produced, but there were three different addresses on the tube. Penilux's manufacturer is a company from Aprelevka outside Moscow, and as you can see on Google Street View, its headquarters are hidden behind a tall gate. Meanwhile, the ingredients are prepared in a grim, red-and-yellow, sixteen-storey block on a typical suburban housing estate. It's hard to imagine a laboratory in Flat No. 6. But who knows, perhaps that's where steel is tempered? And, if the need should arise, comments about Penilux can be sent to an apartment in central Moscow.

Just next to the intimate gel was its bashful counterpart, as pale as a eunuch – Бархатные Ручки, "Velvet Hands" cream made with olive oil and peach kernels. These fans of outdoor gratification really look after themselves. I find a lot of Russian rubbish on my holidays. Reading the addresses on the packaging and calculating the distances makes my head spin. Modern rubbish knows no borders. It travels at lightning speed, on ocean currents, on the wind, and in planes, trucks, articulated lorries and ships.

In the summer heat, rubbish multiplies like bacteria, especially at the seaside. All those dreadful beach resorts, offering piles of single-use junk from makeshift tents and stands. Magnets, personalised pens, cuddly toy seals, balloons with LED lights, inflatable flamingos, bananas and dolphins. There are no toilets here on Dębki beach so the dunes perform that function. You can always find viscose tissues, toilet paper and bin bags torn open by foxes. At lunchtime the beach empties, and people only start to gather again in the evening. To talk about life, politics and work. To drink alcohol, laugh loudly, to sing campfire songs. To use Penilux.

2

A BRIEF HISTORY OF RUBBISH

In the beginning there was rock. I'm searching for the oldest Polish waste: pieces of flint chipped away by Neolithic miners, remnants of tools or perhaps the tools themselves. I'm just a few hundred metres from the edge of the Krzemionki reserve near Ostrowiec Świętokrzyski, the best-preserved prehistoric mining site in the world. The grounds are closed to visitors due to ongoing renovation work on the underground route through the mineshafts, but in the museum building you can see a small exhibition, the most interesting part of which is a short film about the praying mantises that live here. There's only enough entertainment to fill around fifteen minutes, so I've decided to go on a prehistoric treasure hunt. An early December dusk is falling and I'm wandering around the edge of a field, unsure what it is I'm trying to find.

I was expecting an industrial landscape from five thousand years ago, but all I can see are the melancholic hills of Świętokrzyskie province. I was hoping for a lunar panorama of ancient shafts and pits, the remains of Neolithic stonemasons' workshops, but instead I'm confronted with a young pine forest. Beyond it lie undulating lines of ploughed land, with forlorn pear trees along the field boundaries. Perhaps I just don't know what to look for. Somewhere, a black woodpecker caws and a flock of corn buntings fly past, frightened away from the field. I hear their calls – like a child playing with a piece of bubble wrap. Bird watching is going well, but I'm having less luck identifying signs of prehistoric human existence. I pick up a few pieces of flint as souvenirs, trying to imagine they were once touched by a Stone Age miner.

But I know the mines are here. I'm standing somewhere at the edge of the excavation area that's visible in photographs taken by Lidar, an airborne laser scanner used for designing pipelines and roads as well as for mapping archaeological sites. Millennia have passed. The mining landscape has been levelled by farmers' harrows, buried by leaves and overgrown with moss. It's hidden beneath the forest undergrowth that now covers the surface of the site. The shafts are still faintly visible, resembling filled trenches or caved-in cellars. This historical monument has none of the gravity of Egyptian pyramids, particularly since it's strewn with modern Polish rubbish: bits of wooden fencing, polystyrene sheeting and a large, red bumper, patriotically complementing the white birch trunks. There are also wild boar tracks. The leaf litter has been dug up not in search of flint axes but for fat grubs and roots.

*

"Waste means any substance or object which the holder discards or intends or is required to discard."[29] That word, "waste", is so cold and technical. A kind of euphemism that puts a safe, aseptic language barrier in place, cutting us off from the essence of the thing. Rubbish is after all something commonplace and close – ultimately, is there anything more familiar than "old rubbish"? Something homely, warm, almost corporeal. Physiological. In the end, products travel through the digestive system of our daily lives: through the mouths of our wallets, the stomachs of our homes, until we finally throw them out, into the bin. We buy, we use, we excrete. We excrete rubbish.

Though let's not overstate that closeness. Waste also represents a kind of parallel universe, the antithesis of our own, more orderly one. That's why we can't stand having it around us: we subconsciously sense that it brings chaos and disease. Waste is something shameful, dirty and dark, cast out of our minds and our homes. Outside, on the street, it doesn't bother us so much, we've grown used to its ubiquity – because when you take a closer look, you realise it's literally everywhere. Amassed waste is unsettling, like Hitchcock's birds, it feels as if it's plotting a coup, waiting for one false move that would allow it to over-run our world.

Or perhaps this is already happening? What if we looked for rubbish further afield, beyond what our households produce day in, day out? Our civilisation's waste also includes the pesticides in our fields, the periodic table of toxic elements poisoning our rivers (and later oceans), and, of course, carbon dioxide. Safely stored underground for millions of years in coal-seams or crude oil, it's now feverishly pumped into the atmosphere by our voracious demand for energy. What is the filth that's belched out from power station chimneys or engine exhaust pipes if not our civilisation's rubbish? Of course, the chimney expels waste, and the wind carries it far away. But, like household refuse, it doesn't just disappear.

*

Waste is defined as something useless, something to be thrown away. Nowadays we habitually dispose of clothes, furniture, appliances or books in order to make space for new purchases, but this wasn't always the case. In the past, objects had greater value and were made to last longer. People weren't quick to discard things – much of what was left behind by ancient civilisations is real, genuine rubbish. Irredeemable rubbish, like smashed dishes and broken tools, which couldn't be used for anything else. For centuries, man utilised almost anything he could get his hands on. Hunted animals were consumed almost in their entirety. The meat was eaten, the hide used to make clothes, the bones fashioned into tools.

When humans had to move, they took with them whatever they could use. If they were in a hurry, fleeing enemies, they'd take as much as they could carry or would fit on a cart. If they were moving slowly, without urgency, because the soil was turning barren or the groundwater level was rising, they'd gather every last trinket, even taking their huts apart for the journey. That's why, in many cases, nearly all that's left of ancient settlements are the outlines of their foundations.

*

The oldest Polish rubbish may well be waste from the work of the Krzemionki miners, but knowledge of the prehistoric mines at the foot of the Świętokrzyskie mountains was lost for almost four thousand years. The land was taken over by forests. It was the geologist Jan Samsonowicz who restored Krzemionki's place in history, when in the summer of 1922 he ventured inside a quarry that operated there. He found remains of old shafts, pits and mining tools, and traces of the torches used to illuminate the walls. Archaeological circles quickly realised the importance of the discovery. In the late 1920s, the land was bought up and taken over by the National Museum of Archaeology.

Years of research in the excavation area have revealed the remains of around four thousand mines. Extraction took place here between 3900 and 1600 BC. The development of mining techniques can be tracked from the most primitive pits to chamber mines that remained in use for hundreds of years. The distance between the shaft entrance and the mine face could be up to 20 metres – hewing that far into calcareous rock at a depth of 9 metres was the work of generations. In the site's heyday, more than a dozen people would have worked in a single mine, true mining clans providing raw materials to specialised producers of flint tools.

Striped flint, with its alternating streaks of whites, browns, blacks and greys, an unfailingly beautiful, unique rock formed by the warm, shallow seas of the late Jurassic. It's no wonder that at the height of its popularity, not just tools but also ornaments made of this flint travelled far and wide – mass production and international trade in the third millennium BC. It's remarkable to think that axes from Krzemionki have been found in what is now Germany, the Czech Republic, Slovakia, Ukraine, Belarus and Lithuania, in some cases, more than six hundred kilometres from the deposit and the settlement where they were produced.

The spoil tips around the shafts, that is, the fragments of chalky rock discarded at the surface, could be as much as several dozen metres in diameter. The tallest of them were 2 metres high. Mine dumps were formed where the flint was split into smaller fragments. This was the first man-made waste on Polish soil, the rubbish of people belonging to three distinct archaeological cultures: the Funnelbeaker culture, the Globular Amphora culture and the Mierzanowice culture. The huge-scale extraction in Krzemionki, over thousands of years, has irreversibly changed the landscape. A new forest grew over the post-mining waste, while the exposed limestone provided good conditions for stenothermic plant and invertebrate species such as praying mantises and southern snails.[30]

*

Eventually, bronze replaced flint. The people of the Mierzanowice culture, who by then were only producing tools on a local scale, snuffed out the torches in the mine corridors. But even when new, better materials were found, objects lived on for hundreds of years, their form and purpose altered as time passed. In Eastern Ukraine, I heard about Cuman statues in the steppe being used to prop up village fences or carved into troughs for animals. In the same spirit, a miners' stone hammer might have a second life propping a henhouse door. A later, rather shocking example of such thrift are the Jewish headstones used to build roads, kerbs or stairs, like the ones a friend showed me in the backyard of a house on Halytska Square in Lviv.

The Metal Ages heralded the true birth of recycling. Bronze ornaments were melted down time and again into new, fashionable designs. Weapons were reforged into tools and vice versa, depending on what was needed. Anything that could be repeatedly transformed lasted for centuries – like silver Arab dirhams that were treated in Poland as raw materials for smelting jewellery. How did they get here? The Piast dynasty traded with the Arab world; caravans brought silver in exchange for wood, slaves and weapons. Precious metals and stocks of treasure served as payment for men-at-arms.

By 2012, around thirty-seven thousand Arab coins had been found in one hundred and sixty hoards across Poland. Silver really started to flow from the south in the early tenth century (when the Piast dynasty began), but it had found its way here long before then. Coins were not only acquired through direct trading relationships, they were also brought to Polish lands by the Vikings, who had been trading with the Arabs since as early as the eighth century.[31] Dirhams were established as the local currency, which the foreigners from Scandinavia exchanged with the inhabitants of Pomerania for amber, slaves, grain and honey. In the ninth century these silver coins started to be worn as ornaments, too. As fashions changed, they were melted down and reforged so many times over the centuries that all trace of their origins disappeared. A world in which extracting raw materials was costly and labour-intensive took great care of its resources.

Recycling expensive metals was a matter of course. Even today, nobody really throws out gold earrings. But it wasn't so long ago that nearly every item enjoyed a long and fruitful life. In her excellent book, *Waste and Want: A Social History of Trash*, Susan Strasser describes the case of an American businessman, a certain Morillo Noyes, who in the mid-nineteenth century traded in absolutely everything and anything that could be recovered and transformed. At the peak of his career, he led a twenty-two-strong division of peddlers whom he supplied with a horse and cart from his own stable. He bought, sold and bartered, and was constantly on the lookout for new opportunities. When he heard about a company producing dyes from old leather, he set off the next day to see the manufacturer, one Mr Dodd of Newark. During his visit, he noted that the site purchased old shoes, 35 cents for 100 lbs. They had to be relatively clean and made of leather. Dodd also bought horns and hoofs.[32] Literally everything had some kind of value. Back then, the world produced almost no waste.

Going from door to door, Noyes' peddlers sold a wide variety of essential products, such as needles, rolls of material or tin bowls. The business also traded wholesale, buying directly from factories and selling to shops. Items acquired from individuals were sold to foundries to be recast. Purchasing prices were fixed. The peddlers were interested, among other things, in beeswax, feathers and pig bristles as well as deer, sheep, raccoon and even cat skins. They bought horns and hoofs for paper mills, which used them to produce gelatine for sizing paper (so that the ink didn't run), and fat, which could be used as a lubricant or as fuel for lamps.[33]

As recently as the nineteenth century in the United States, that remote, enormous yet provincial country, items were in principle repaired indefinitely. Old clothes were salvaged in many ways. A good housewife would replace material where it was worn. Cuffs and collars that couldn't be thoroughly cleaned were turned inside out. Servants wore their master's old clothes. Only the wealthiest people had several changes of underwear. Some households even used untreated urine to clean equipment, poured straight from chamber pots each morning.[34]

As time passed, such puritanical modesty, thrift and frugality remained highly valued, but the country was growing in wealth and power. Wave after wave of immigrants were arriving from Europe, and industry was booming as a result. And with industrialisation came the true beginnings of mass production and thereby of mass waste. By 1897, the department store Sears' catalogue was seven hundred and eighty-six pages long, not including twenty-four detailed leaflets focused on specific sectors.[35] Between 1899 and 1927, the productivity of American industry increased fourfold.

In the early twentieth century, many products in the United States were already sold in packaging, but it was packaging that could be reused many times over. Tobacco tins were designed so that they could later carry sandwiches. People therefore had to be patiently persuaded that throwing things out was not a sin. That's right – it was a skill people had to acquire. In other parts of the world, it was a trickier lesson to learn. Thirty years ago, inhabitants of Eastern Europe were still carefully holding on to items that Americans had been nonchalantly throwing away for decades. Indeed, it was quite common for people to put collections of Western cosmetics packaging or beer cans on display in their homes.

I remember how, twenty-five years ago, in a spirit of stubborn wartime and post-war prudence, my grandmother used to keep hold of utterly worthless bits of packaging. We used to laugh at her. When everyone was obediently learning to throw things out, she patiently washed cartons and never disposed of anything too hastily. Born before the First World War in the Minsk Governorate, my grandmother never found out that she was a trendsetter in Poland, embodying the principles of zero waste and upcycling before anyone had even thought of those names.

A breakthrough and rapid acceleration in the art of throwing things out came with increased awareness of the existence of microorganisms. The history of hygiene is, by the way, an example of how twisted the paths of progress are. The high standards of antiquity were abandoned and forgotten in Europe for many centuries. Frankly, when they conquered Rome, the barbarians plunged us into profound hygienic ignorance. The

Visigoths destroyed bath houses, believing that care for cleanliness made men soft. The Germanic peoples styled their hair with rancid butter. I don't know which would have been more terrifying: their contempt for death or the stench they spread. The rise of Christianity did nothing to improve the situation. Its focus on spirituality, the belief that all bodily concerns led to sin, was another hammer blow for hygiene. A meeting with Saint Jerome must have been tough on the senses, since he believed that after baptism washing was no longer necessary. How did Saint Francis form such close ties with animals? Perhaps it was his *olor de santidad*, his odour of sanctity, that aroused their interest.

As Katherine Ashenburg writes in *The Dirt on Clean*, "Rich and poor, men and women lived in close connection with each other's dirt, excrement and bad smells."[36] The situation slowly began to improve in the eighteenth century, though it would still take years for societies to completely abandon certain superstitions and habits. People were afraid of catching a cold, so they wouldn't change their clothes for years. At most they'd wash their hands and faces, because they feared the warm water would open their pores and allow diseases to be absorbed through their skin. It was only after Pasteur and Koch's research led to widespread awareness of bacteria and the transmission of infectious diseases that good hygiene became standard practice. Before then, it could in no way be taken for granted that doctors or medical staff would wash their hands after hospital procedures.

A true revolution came in the 1920s, when, aided by the

development of advertising, hygiene became big business. The ideal world thus became one that was sterile and odourless. Our natural bodily scents were increasingly shameful, associated with dirt, poverty and neglect. Spreading a stench was seen as a brand of antisocial behaviour that could be remedied, for instance, by using Listerine mouthwash or Odorono, the first deodorant. Taboos about the human body were superseded by intolerance of unpleasant odours. And there was something that was far more shameful than the smell of your armpits: the matter of menstruation.

Few products were truly single-use before Kotex sanitary pads appeared on the market in the early 1920s. Before then, women had used washable cotton cloths. The name Kotex came from the abbreviation of "cotton-like texture", as it was produced from an absorbent material that had been used to make bandages during the First World War. The company found a new purpose for the surplus stock lying in their warehouses, and adverts for sanitary pads started appearing in popular magazines, a watershed for an age-old taboo. Of course, the copy avoided any direct reference to it, the message was conveyed clearly enough by the claim that it "catered to women". The company was able to persuade pharmacists to display the product on their shelves, and by 1922, pads were already being sold in vending machines in public toilets.

Market research showed that women were embarrassed to leave tissues soiled with menstrual blood in the bathroom. They wanted a product they could flush down the toilet. So

Kotex seized on this as the most important marketing angle. The pads' packaging included instructions on how to dispose of them: remove the layer of gauze, tear through the filling then wait a few minutes for the pad to absorb water. But the whole procedure took too long, and many customers complained that it was no more hygienic than dealing with the old cloth towels. The new pads also had a habit of clogging up pipes.

Strasser's book clearly shows how our modern consumer culture originates from American inventions of the 1920s. Even then, companies producing sanitary pads were already competing among themselves, commissioning extensive market research and delegating advertising to specialist agencies. Kotex, initially targeted at upper-class women (as the pads were quite expensive), gradually shifted towards the working classes, and, above all, to immigrant families. Opting for single-use pads was meant to be seen as a status symbol, an aspirational step towards a new, better way of life.

*

The door to throwaway culture was open. Companies strived to ensure that just one single use left their products fit only for the dustbin. Like paper cups, for example. But it wasn't just about one-offs, it was about using everything more fleetingly and regularly buying new models. Adverts declared that holding on to old items was what poor people did. The market was filled with new products that were supposed to work better, but which in fact differed only in appearance. They were simply newer and more fashionable.

A major turning point arose from the rivalries between the big automotive companies. The car was, after all, the key symbol of what we recognise as the American way of life. From 1923, General Motors began releasing a new model of car each year. It was a radically different strategy to the one Henry Ford promoted in his company, which had the biggest market share at the time. Driven by sincere principles, Ford stubbornly stood by the belief that the ideal car should be so good that nobody would want to buy a new one. General Motors, meanwhile, tried to stimulate a desire for novelty, or rather, a feeling of dissatisfaction among consumers. In spring 1927, one of GM's brands – the Chevrolet – overtook sales of the famous Model T Ford.[37]

After the initial surge of mass production, many sectors of the market started to slow. If growth were to be maintained, consumers had to be encouraged to buy. At some point in the late 1920s or early 30s, Christine and J. George Frederick started using the term "progressive obsolescence". This was the name they gave to a novel consumer trend exemplified by a passion for all things new, a readiness to spend money on them and a tendency to throw out old products, even if they still worked, in order to buy more stuff. This fresh approach to life was declared typically American, a contrast to boring, old-fashioned Europe, which was still fascinated with antiques and where everyone had shelves of dusty keepsakes at home. Christine and J. George were later called the "evangelists of the new ideology", spreading a cult of newness and an aversion to all things old and unfashionable.

Interestingly, only a few decades earlier, Christine Frederick had been an advocate of thrift, common sense and economy. We can see that in her book, *The New Housekeeping: Efficiency Studies in Home Management*, published in 1913, in which she shares dozens of tips for saving time and money in the home. There's something obsessive in her optimisation of every activity. Frederick reported that she saved fifteen minutes thanks to a new way of organising the washing up, for instance by thoroughly clearing plates of leftover food. With badly arranged equipment, work in the kitchen had one "crossing and recrossing like the tracks of a hound after a hare",[38] she admonished her readers. The book contains diagrams explaining, for example, how to arrange a kitchen to save a few steps when fetching a pan from a cupboard. In 1913, she advised that one ask oneself before buying new kitchen equipment: "Do I need it? Will I get my investment out of it?". Back then, the overriding value was economy. Economy of fuel, economy of time, economy of effort, even down to a few steps.

Just over a decade later, it was all about keeping the market at full tilt, rather than promoting individual rationalism. The "evangelists of the new ideology" began by reforming language. Purchasing new products, equipment or devices when the old ones still worked had been associated with excess and extravagance. Henceforth, the sorry terms "obsolete" or "out of date" began to appear alongside references to progress and innovation. Throwing out old devices was given the joyful name of "creative waste". Advertisements stressed the need to keep up with the times, to follow new trends. Consumers competed with one another to be the first to get their hands on the latest novelties. Modernisation and constant movement were the cure for stagnation.[39]

The concept of "progressive obsolescence" spread virulently. Using old appliances was branded as resisting progress, a betrayal of society, which grew wealthier as the economy flourished. Even during the Great Depression, sales of expensive appliances such as fridges actually increased, thanks to dazzling adverts, which claimed that the fridge was worthwhile because it kept food fresh for longer. People sewed clothes out of cotton sacks but purchased new coolers.[40] World War Two, a time of sacrifice and renunciation, interrupted this sustained, methodical lesson in buying and discarding. Rubber, old silk and paper were collected, while cellophane and tin and steel products vanished from the market. People used to say that an old bucket was three bayonets, an iron was two helmets, and two pounds of fat from the kitchen was enough glycerine for five anti-tank shells.[41]

But, naturally, once the war was over, people no longer wanted to deny themselves anything at all.

*

"Plastic" doesn't always mean the same thing. It's a collective term for thousands of different polymers, making up practically everything around us. It comes from the Greek, *plastikos* – meaning malleable, fit for moulding. The birth of plastic is considered to be 1839, when Charles Goodyear first made vulcanised rubber. But humanity had natural polymers long before this. One example is shellac, the resin secreted by insects that feed on the sap of certain trees, and which was used for centuries as a dye and an ingredient in varnishes and lacquers. Or Gutta-percha – sap from the tree of the same name, which, when cooled, becomes a hard, resilient substance, used in Indonesia for making machete hilts. In the mid-nineteenth century, it was used as insulation to protect underwater tele-graph cables.

The next essential step in the evolution of plastics was work on cellulose-based materials, leading to the creation of celluloid, the first widely used thermoplastic, in the 1870s. However, a number of key drawbacks prevented its extensive application: celluloid is highly flammable and deteriorates when exposed to light.

Early plastics were above all intended to replace expensive, difficult-to-obtain, natural materials. A substitute for ivory was in particularly high demand. Phelan & Collender, a company that manufactured billiard balls, even offered a reward of ten thousand dollars to the inventor of a suitable replacement.

The first mass-produced plastic was Bakelite. It was patented in 1907 by Leo Baekeland. The basis of this material was a phenyl formaldehyde resin, making it the first material made of ingredients that did not occur in nature but were created synthetically in a laboratory. Bakelite was labelled the material "with a thousand uses". Its isolating properties were without doubt key to its success: it was resilient, non-flammable, a poor conductor of heat and electricity and resistant to atmospheric conditions. In 1925, the Thermos company made a beautiful brown thermos flask out of Bakelite, and a few years later Siemens' Bakelite hairdryer appeared in shops.

Plastics were liberated from the role of substitutes in the 1930s. The economic crisis and heightened competition created the right conditions for industrial design to flourish, and ever more versatile plastics provided endless opportunities for innovators. One of them, Harold Van Doren, wrote in 1936: "Six or seven years ago, the widespread use of plastics was unheard of. The phenomenal increase in their employment [. . .] can be attributed to a number of factors. For one thing, the availability of brilliant colour and pastel shades. Durability of surface, accuracy of dimension, extreme lightness. Perhaps, most of all, the desire for something new, something different."[42]

Roland Barthes began his essay "Plastic" from the mid-1950s with the somewhat disappointing observation that the names of plastics sounded like unknown Greek shepherds: Polystyrene, Polyvinyl, Polyethylene. Some of his reflections have become very dated, while others remain strikingly perceptive. It's not correct that plastic is "best revealed by the sound it gives, at once hollow and flat", nor is it the case that plastic is "powerless ever to achieve the triumphant smoothness of Nature". The truth is that for plastic, nothing is impossible. It is a material that will replace others, because – and here Barthes' prophecy was not mistaken – in reality, "plastic" is an infinite number of combinations.

When Barthes published his essay, plastic wasn't yet so over-whelmingly ubiquitous. Chairs were wooden, milk was sold in glass bottles, and cars were still just piles of scrap metal. Plastic items that are completely commonplace nowadays were at that time the height of fashion, masterpieces of industrial design. Like the polyethylene bucket with a metal handle, given the technical name KS 1146 and produced in 1954 according to a design by Gino Colombini. Or Robert Menghi's beautiful silvery-green polyethylene gas canister. Barthes associated plastic with fakes, imitations or junk: artificial diamonds, silk and fur; he held it in contempt, insisting that it lacked charm. "Lost between the effusiveness of rubber and the flat hardness of metal", it was hard to grasp, escaping all categorisation. But his diagnosis was accurate. "The hierarchy of substances is abolished: a single one replaces them all: the whole world can be plasticised, and even life itself, since, we are told, they are beginning to make plastic aortas."[43]

I think Barthes would be amazed at how quickly we've transformed our world into a plastic empire. We are at once its beneficiaries, its victims and its hostages. Our world cannot exist without plastic, but we can't deal with it and nor can nature. Turtles eat carrier bags because they can't distinguish them from jellyfish. The structures created in laboratories a few decades ago are so deceptively similar to natural substances they have become traps that evolution is unable to outsmart. And we, the perpetrators of this confusion, are helpless.

My friend's parents bought a holiday flat, inheriting a plant that the previous owner had left behind on the balcony. They spent barely two weeks a year there, but by some miracle, the plant didn't die. It was some kind of undemanding, southern species, with leathery, light green leaves. When those finally began to fall off, the owners breathed a sigh of relief – the plant was at last behaving like a living organism. They'd been unnerved by its logic-defying determination to live. When they cleared the leaves up, it turned out they were made of soft plastic. They had simply come unstuck. It was a replica of the highest quality.

*

Single-use items came of age in the 1950s, aided by a booming plastics industry. This is well illustrated by a photo from the August 1955 issue of *Life Magazine*.[44] It depicts the ideal American family: a blond man with his blonde wife and their blonde daughter. But there's something strange about this tableau. Something asymmetrical. Just one child? Surely they need a little boy to complete their bliss. Nowadays there'd definitely be a determined-looking dog, too. A Labrador or perhaps a Border Collie. Regardless, the three of them are slim, healthy, young and very happy. Frozen in a joyfully carefree moment, they're tossing all manner of plastic plates, mugs, cups, buckets, trays, bowls, knives, forks and spoons into the air, along with some other pieces that are difficult to identify in the small, black and white copy on the internet.

It's an eloquent illustration of the safe, warm and homely 1950s. A dream of sleepy suburbs, drive-in cinemas and the gradual enrichment of white Americans. Not a word about those pesky communists who want to set the world on fire, about Cuba or the war in Korea, or visions of nuclear apocalypse. The family are flinging their arms open wide, a weird, rather mannered pose for an outburst of joy. It looks more like the expansive gesture of a priest, an imitation of the statue of Christ the Redeemer in Rio de Janeiro, or the typical stance that he adopts when feeding the five thousand. The family is creating a new world, one of abundance in which "no housewife has to worry about washing up". The plastic enthusiasts of the time had carefully calculated that cleaning all this plastic crap would somehow take forty hours.

I'm not buying it. It might take a five-year-old that long, but certainly not your stereotypical Mary-Lou, champion of kitchen gymnastics. But the advert isn't supposed to tell you the truth: it's about selling dreams. The dream of free time for yourself and your family. Washing up is boring – it may take twenty minutes or ninety, but it feels like forty hours. This brand of nonsense was the harbinger of modern shopping channels that bravely stand alongside housewives in the fight against daily dirt and disorder. The idea of forty hours of washing up is as alarming as the image of a housewife wiping beads of sweat from her forehead as she frantically scrubs encrusted grease from the cooker. Relax. We know they're about to offer us a miracle cure.

"The objects [. . .] in this picture [. . .] are all meant to be thrown away after use," wrote the editors of *Life*, just to be very clear, in case some browbeaten grandma actually wanted to wash it all. From then on, "throwaway living" was a mandatory way of life. Sixty years later, this sweet picture looks grotesque. In the tape-machine of my brain I press <<REW and a looped, never-ending shower of plastic rubbish rains down on the cute, laughing family. But they're still happy. There's another telling photograph from the same series that shows the crew diligently clearing up after the photoshoot. One man has even taken off his shoes to avoid soiling the sheet spread over the set, his comfy, worn-out sock twinkling in our direction. In the end, somebody did have to tidy up that mess.

*

We've had our fill of plastic and we know it's done us no good. Consumer culture has taught us that we have the right to choose. We want to make informed, thought-out decisions, but the world is too complex for us to be experts on everything. Illusory help is at hand on the internet, the Great Dustbin, where valuable knowledge is presented on an equal footing with gibberish. Conscious and calculated disinformation is commonplace, like in the case of "eco-friendly" plastics. Every few days, I read about plastics that are supposedly natural and biodegradable, properties that are now so coveted by consumers. But just because something was produced from plants doesn't automatically mean it will obediently decompose as readily as a tomato skin. And, indeed, the opposite is true: even materials made from oil can be biodegradable. Crude oil, incidentally, has completely natural origins, it's nothing more than the remains of plants and animals.

I've heard of plastics made from crustacean shells, from mushrooms, from various natural fibres. Of course, none fit the bill perfectly. Some aren't elastic enough, others react badly to humidity. Some, of course, do biodegrade, but that happens to be their only quality. Many "eco-friendly" plastics decompose in exactly the same way as the oil-based polyolefins we are trying so hard to escape. Recently, our saviour was supposed to have arrived in the form of polylactide (PLA), a plastic produced from corn meal.

The enthusiastic, almost ecstatic articles in the press, which sounded more like sponsored features, led us to believe we were on the brink of a revolution. The manufacturers assured us that polylactide meant no "chemicals" and no harmful substances – just pure nature. As if it grew on trees. Polylactide is a polymer, which is created by carrying out a series of chemical processes: hydrolysis of plant waste followed by fermentation with the aid of bacteria. The lactic acid produced this way is treated at differing temperatures with acids, bases and solvents. It's claimed that polylactide is biodegradable, but the articles fail to mention the awkward fact that decomposition only takes place in conditions of controlled temperature and humidity. In Poland there are no sites capable of implementing this process within the necessary six-week period.

According to simulations by a polylactide manufacturer,[45] in landfill conditions with no access to light or air, the material takes up to one hundred years to decompose, slowly breaking down into microplastics. In this respect it's essentially no different from traditional polyolefins. And there's another problem with bioplastics. Materials such as polylactide do not currently have separate collection schemes or a marking system that would allow them to be removed from conventional plastic waste streams. This is a major problem when it comes to achieving a "closed-loop" system. What's more, the pollution of polyolefins with plastics like PLA significantly reduces the quality of the recyclate (the granulated plastic created during the recycling process).

*

I push a fragment of a thin, transparent structure across the table.

"What's this?" I ask deviously, because I already know the answer.

Michał, a chemical engineer and polymer specialist who has spent months patiently explaining the nature and properties of plastics to me, listening to my quandaries and helping me resolve them, frowns and picks up the object. He touches it, bends it around for a while, and then speaks carefully.

"It looks like slightly yellowed, degraded polystyrene. It's lost elasticity and become brittle."

I've never been very good at sustaining tension, and much as I'd like to keep pulling his leg, I burst out laughing. Michał looks at me tetchily, realising he's been tricked.

"It's a squid pen," I explain, feeling very proud of myself.

I only know that because I saw it being filleted with my own eyes.

Isn't it odd? Could we go so far as to say that plastic has become more familiar to us than a squid pen? In the last few months, I've started seeing it everywhere. One dark day in February I was trying to spot a wren amid a tangle of branches in the park. It was chattering away, irritated by my presence. Its feathers were smooth and glossy like polyester. Or take the pintail, with its shimmering polished beak that looks as if it's always damp, as if nature had prudently made it out of rubber. That's what I mean when I say I see plastics every day. Everywhere. We've grown so used to their ubiquity that they seem to go without saying, they're manifest, all around us.

We see polystyrene in various forms on a daily basis, whereas we rarely see a squid pen, if ever. And yet, as a species, we've only come so far because we were receptive and observed nature carefully. We still do. Military robots look like animals. High-speed trains mirror the silhouette of a kingfisher. Even high-performance gearboxes imitate the mechanism by which fleas hop. This is known as biomimetics. In the end, the solutions adopted through patient evolution turn out to work best. Solutions that have been tried and tested for millennia.

Michał takes a lighter and ignites the pen. It's the simplest, albeit fundamentally flawed, way of checking what sort of material we have at hand. The chitin catches easily, though what matters isn't the fact that it ignites, but how it burns and what odour the smoke gives off. It smells of singed hair.

"Polystyrene would have a flowery scent, this smells like something organic," he explains, and quickly adds, "but on the other hand, polyamide would also smell like that. If you hadn't said it was a squid pen, I'd have gone for polyamide."

FEBRUARY 4, 2020

10.48 a.m., 52°12'58.0"N, 20°59'01.5"E

JULY 23, 2018

6.18 p.m., 50°47'16.8"N 15°32'13.4"E

In summer, the grass was orange and white. Drivers would stick their hands out of the window and lazily flick the ash off their cigarettes – but not too lazily, as the heat was unbearable. The smokers preferred to suffer nicotine cravings rather than exposing themselves to the glare of the scorched city. Before winding the window back up, they'd drop the butt through the gap onto the verge. It's strewn with cigarettes because there are always cars idling here, at the foot of the traffic lights next to the Oncology Department – is that not an irony of fate?

Nonetheless, cigarette butts are somehow less of an eyesore here. After all, they're a permanent feature of urban landscapes. They're more offensive in places where we expect to find untouched nature. Outside the "Łabski Peak" mountain refuge in the Karkonosze National Park, it only took me a minute to gather a decent pile of butts – of men's cigarettes and women's, thick ones, thin ones, some smeared with lipstick, others barely lit and still more smoked greedily, right down to the orange paper.

The smokers were standing less than ten metres away from a bin.

How many times have I too casually thrown my finished cigarette into some shrubs or a lake? One flick . . . It's such a delightfully nonchalant move. Like in a film. It's what everybody does. And besides, what's wrong with one little ciggie? Not much, I told myself, but that was before I grasped the true scale of the problem. Every year, six trillion cigarettes are produced globally.[46] Smokers drop at least two thirds of them wherever they please. Cigarette butts, or filters to be precise, are thus the most common type of rubbish in the world. A report by the Ocean Conservancy group, which organises beach litter-picks, states that in 2019 its volunteers collected around 5.7 million cigarettes. Next in the rankings came food packaging

(for sweets or crisps) and plastic straws.[47] According to studies in the US, cigarette butts make up more than a third of all rubbish found on roadsides.

Nearly 90 per cent of cigarettes contain filters made from cellulose acetate, a polymer that is also used to make frames for spectacles. Depending on conditions, a filter can take between a few years and several decades to decompose.[48] But apart from the filters, there's the smoke, which contains dozens of toxic and carcinogenic compounds. Many chemicals in cigarettes are found in other well-known products, but in those cases they're properly labelled. Ammonia is used in household cleaning products, arsenic in rat poison, while nicotine is used as an insecticide. All these substances pollute soil and water.

For a while I took solace from an article in *Nature* that described how titmice living in towns use cigarette butts to build their nests. The chemical substances in tobacco were said to act as a repellent against mites.[49] I saw this as comforting proof that "nature will cope". I owed my optimism to research by scientists from the National Autonomous University of Mexico, who had studied sparrows' nests and house finches.[50] But two years later, the same team found that contact with toxic substances (nicotine, ethylphenol, titanium dioxide, propylene glycol, insecticides and cyanide) had damaged the DNA and chromosomes of chicks and adult birds.

Poison then, after all.

3

THE GANNET'S PREDICAMENT

It's like a dream. It's October and yet there are still robins singing in the green gardens. The warm sea bubbles like prosecco and we're listening to "*E penso a te*" by Lucio Battisti on repeat. We don't understand much, but somehow we know it's a song about the beauty of life. The boat rocks gently, the boxed wine tastes incredible. At moments like this I'm suddenly racked with guilt. Supermarket cashiers are freezing in the dark as they wait for a bus, men are taking furtive sips of beer in the tram. Back in my country, all is grey, damp and drowned in smog.

Each island of the Lipari archipelago is different. The air in Vulcano is always heavy with the smell of sulphur. Mud from the hot springs leaves a terrible stink on the skin, seeping into every fibre of your clothing. A huge crater and smoking expanses of yellowing rock. An empty, silent landscape. This is what the world looked like millions of years ago. Somewhere on the slope some chaffinches take flight in silence, as if overawed by the church-like atmosphere.

As for Panarea, she's more like a lesser goddess or a green nymph. Haughty and inaccessible like her ruddy, rocky cliffs.

Up here, looking out to sea as it wrinkles like bright-blue tissue paper, watching the changing light searing through the clouds, it's clear why mythical stories set their scene in places like this.

Stromboli: a huge cone, the archetype, a child's drawing of a volcano. Perfectly symmetrical with a sliced-off peak. It's a gentle climb, slowly winding upwards; it feels like we've barely left the foaming sea behind, yet we've already climbed nearly a thousand metres. Every now and then we hear a stifled rumble. The entanglements of African reeds disappear at the top, leaving just the black slopes of volcanic sand and the light of the sun setting somewhere on the other side of the mountain.

The sun is very low when we reach the summit and the craters are smoking at our feet. They're like siblings, each with their own personality, character and temperament: sanguine, melancholic, choleric and phlegmatic. Silently and with the unwavering rhythm of a steam engine, the most hard-working of them tirelessly spouts little clouds of smoke. Puff-puff-puff-puff. The largest one, slightly further back, commands respect. Quite regularly, with a loud, heartfelt moan, it spits a plume of lava, and the light volcanic rock drops with a hollow clatter to the crater floor. But most forbidding of all is the small hothead, slight and narrow, torn apart by a consumptive cough that turns into the roar of a jet aircraft at take-off. Like mortar fire, it launches rocks a few hundred metres into the air. Darkness falls, it becomes cold, but down there, life goes on uninterrupted. The lava bubbles like Bosch's hell. And then, suddenly, when we turn to go, as if it's saying, "Leaving already? But we haven't had a chance to talk," a hidden crater explodes on the right-hand side. It erupts just once, but properly. Low and loud. As if the Tsar Cannon imprisoned in the grounds of the Kremlin had finally spoken. Volcanoes give you a kind of soothing sense that the Earth hasn't said its last word and that forces exist that are more powerful than mere money.

And so the days pass. One evening we're wandering aimlessly around a small town when an elderly man we talk to in the street invites us to his restaurant. The place is called La Cambusa, and judging by the photographs on the walls, some famous faces, albeit ones we don't recognise, have dined there. As soon as we sit down, a sort of diabolic chef-cum-fisherman in a baseball cap and puffer jacket appears and with a flourish presents the star of the show: our dinner. The sea monster grins roguishly back at us and gives us a dull wink. The owner gesticulates wildly, saying something about life perhaps, or about wine, or women – it doesn't matter, we're loving the spectacle, so we agree to everything. The Nero d'Avola practically pours itself.

I feel embarrassed when things are just too good, but maybe I shouldn't be so hard on myself. Perhaps it's worth appreciating life when it's like this. Half an hour later, a bowl of pasta with our barracuda arrives at the table. Top marks for enthusiasm, but the flavours are a little bland. Though let's not spoil the mood, let's savour the moment. As I'm reproaching myself in my head again, my tooth strikes something hard that isn't a fishbone. An unpleasant sensation, like when a bucket you were expecting to be full turns out to be empty.

Whatever it is, it's refusing to give in, so I don't fight it. I spit out a hard, white ball. Some of whatever it's made of has crumbled against my tooth. The little ball is transparent.

*

It's more than twenty years since the sailor Charles J. Moore discovered a huge rubbish dump floating in the ocean between Hawaii and California. A mass of plastic packaging, bottles, lids and countless tons of unidentified waste. The area was named the Great Pacific Garbage Patch and its growth has been monitored apprehensively ever since. New debris accumulates rapidly, carried there on the North Equatorial Current.

A report[51] drawn up in 2018 on the basis of observations from planes and ships estimated its size at 1.6 million square kilometres, at least. That's four to sixteen times larger than previous estimates. The Patch is a floating polymer soup of 1.7 trillion pieces. There's no point trying to illustrate numbers with twelve zeros, if you'll forgive me. It would be like one of those scientific estimates for how many grains of sand there are in the world. The researchers say that theirs is a conservative guess. Where does the rubbish come from? The scientists fished out a few hundred items, on which words or sentences could be identified in nine languages. Among them, 30 per cent of the rubbish had Japanese labels, the same amount again was in Chinese. The rest was traced to nine other countries.

The biggest environmental problem for seas and oceans is abandoned and lost fishing nets. Nets floating on the water never stop doing their job, ensnaring fish, sea mammals and crustaceans. They also catch other rubbish. Produced from highly resilient polyamide or cheaper polypropylene, they don't degrade quickly in marine environments. Like all plastics, they're broken down into small pieces by light, sea water or friction, but this process takes years. In 2018, the mass of rubbish trapped in the Patch was estimated to weigh at least 79 thousand tonnes, of which 46 per cent were pieces of net that had been lost, swept away by the current or just thrown overboard.

While nets and macroplastics make up most of the floating waste in terms of mass, it's tiny, essentially uncountable microplastics that predominate in terms of quantity. Estimates (which are constantly changing) suggest that microplastics constitute more than 90 per cent of all fragments floating there.

The problem of plastic clusters in the ocean is so extensive and enormous that there is already talk of an eighth continent – the Plastisphere. That's a rather misleading image; oceanic rubbish dumps don't have the structure of land, you can't stand on them and trees will never grow there, nor helicopters land. The Great Pacific Garbage Patch is only the most famous of the world's huge marine middens. It's estimated that at least five large vortexes of accumulated plastic have formed around the planet. Most plastics that end up in the sea are less dense than water, which is why they can travel great distances on surface currents or even on the wind. The majority of what is floating around is polyethylene and polypropylene, which reflects their share of global production.

How does plastic end up in the sea? At least 80 per cent gets there from land, mostly on the currents of ten major rivers. Eight of them are in Asia – the Yangzte, Indus, Yellow River, Hai River, Ganges, Pearl River, Amur and Mekong. Two are in Africa – the Nile and the Niger.[52] These rivers and their thousands of tributaries flow through densely populated areas where no refuse collection systems exist. Almost half a billion people live in the Yangzte River Basin, many of whom throw their rubbish straight into the water. The rest of the waste that comes from the land spills into the sea from overflowing surface water, for instance during the great floods of monsoon season. Every day, the internet is inundated with shocking films showing masses of plastic rubbish coursing downstream, chipping away at our conviction that sorting glass bottles and yoghurt pots at home can avert catastrophe.

*

Our individual sacrifices, eco-friendly behaviour and good habits are of course meaningful, but largely just for ourselves. Occasionally I clean up after others and correct my neighbours' recycling, too. So it makes me want to go back to bed when I read that in January 2019, on a journey from Antwerp to Bremerhaven, the container ship *MSC Zoe* lost at least three hundred and forty-five containers carrying all manner of goods, in a storm in the North Sea.[53] Tonnes of products washed up on beaches in the Netherlands and Germany: toys, white goods, electronics, shoes, domestic appliances, car parts. Most containers sank in the region of the Frisian Islands, while twenty were located near Borkum and the same number again were washed ashore.

Two containers carrying hazardous goods were never found. One was transporting almost one and a half tonnes of lithium-ion batteries, the other nearly three hundred cartons of perkadox. That's the general name for a whole group of chemical compounds, in this case a mix of benzoyl peroxide (which in high concentrations damages the respiratory system and is toxic to plants and animals), and dicyclohexyl phtalate (probably a dispersant, which facilitates the mixing of one substance into another, such as pigments into paint). A container carrying over twenty tonnes of polystyrene beads (a granulate used to make polystyrene products) was damaged. Most of the pollution occurred on the beaches of Schiermonnikoog island, barely ten kilometres from continental Netherlands. I won't cite any numbers because they don't hit home anyway. Similar catastrophes occur frequently at sea all over the world. News of lost cargo and its contents rarely makes headlines, as shipowners try to cover their tracks instead. Three hundred and fifty containers lost in the North Sea can't vanish without a trace. But what if the disaster had happened in the middle of the Pacific?

The most famous loss at sea is probably the twelve containers of plastic bath toys that fell overboard in January 1992. The *Ever Laurel* ship was travelling from Hong Kong to Tacoma, but half-way through the journey, near the Aleutian archipelago, it encountered a powerful storm. Waves more than a dozen metres high washed the containers off the deck. The plastic animals (yellow ducklings, red beavers, green frogs and blue turtles) were made in China and are completely hermetic, so they'll never sink and will instead float around the seas and oceans until the material they're made of breaks down.

The first toys washed up in Sitka in Alaska, then in Japan and Indonesia, in Australia and on the coasts of South America. They stopped appearing on beaches in 1996. Curtis Ebbesmeyer, an American oceanographer, saw an opportunity for scientific study amid the obvious environmental concerns. He analysed the routes the toys had taken, proposed a model for the movement of sea currents and formulated the hypothesis that the bathtub armada was heading north. Indeed, at the turn of the century, the ducklings and company started appearing in the Atlantic. The ocean cast them ashore in Canada, Iceland and the United Kingdom.[54] Polish ecologists from a team led by Marcin Węsławski also spotted them at Spitsbergen. It really is no exaggeration to say that waste knows no borders. The oceans are a single body of water, split into different parts purely for our convenience.

Sometimes unexpected natural disasters sweep rubbish off land. In 2011, a tsunami caused by an earthquake devastated the north-eastern coast of Japan and carried off the livelihoods of thousands of people. The most spectacular journey was made by a football belonging to Misaki Murakami from Rikuzentakata in Iwate prefecture. One year and five thousand kilometres later, the ocean deposited it on the shore of Middleton Island near Alaska. It was the first item seized by the tsunami to cross the Pacific. Two years after the disaster, a small fishing boat covered in barnacles was found on the coast of California. Thanks to the markings on its side, it was identified as the property of Takata high school, where Misaki Murakami was a student. The boat had been used during lessons on marine biology.

But it's first-hand accounts that are most striking. Around fifteen years ago, Pau Freixa Terradas, a Catalonian translator of Polish literature and a Gombrowicz expert, met a young geologist on a train in Russia. Sharing a train carriage brings people together, especially in the East, and as the conversation went on, the man invited the tourists to his home and told them about his visit to the Kola Peninsula. He'd travelled there with a group of scientists as part of a *komandirovka*, a Soviet-era research trip. The land was untouched, or so it seemed – there's probably no place on earth more austere than these northern seas surrounded by tundra. But on the shore of the peninsula the scientists made a surprising discovery. The wild beach was strewn with washed-up footballs. Finding traces of Portuguese words on a number of them, the team excitedly seized on the idea that the balls had come from Copacabana. That was theoretically possible – they could have been carried there by the Gulf Stream. I like to picture the group of geologists, struck by the total absurdity of the situation, casting their backpacks to the ground and playing a spontaneous game of football in the midst of that deafening emptiness.

*

In spring, fate decreed that we should meet. As if they'd planned it among themselves, various friends started telling me about Dr Agnieszka Dąbrowska, a researcher at the Intermolecular

Forces Laboratory at the Chemistry Department of Warsaw University, one of the few Polish experts in plastics in the marine environment. Agnieszka, a keen yachtswoman, studies plastic pollution in seas and oceans – in the Antarctic, the Arctic and in Europe. "Isn't it ironic that we've created practically indestructible materials to produce items we only use once?" she asks philosophically. She cast her first metal container for collecting samples into the Mediterranean Sea from on board the *STS Pogoria*, a sailing ship. She didn't expect to find anything. But after just five miles of trawling, tiny fragments of microplastics had collected on the filters. The sample collector Agnieszka uses has a nickel-chrome mesh with holes measuring 20 micrometres in diameter, though other scientists use meshes of varying sizes. There is no single, universal research methodology.

The concentration of plastic in the sea varies significantly depending on where the samples are collected. You'll get different results in Indonesian waters and in the English Channel. The specks are difficult to count because they're present at different depths and move in all planes. And it's hard to identify them. Agnieszka says that studying plastic under the microscope is like listening to the buzz of conversation in a café. It's only when we record that uniform hum with a sensitive device that we can identify specific tracks, hear individual words that we can join up to derive meaning. We can even pick out background noises amid the buzz – the clink of spoons against a cup or pouring sugar. A spectrophotometer is that kind of precise instrument.

Light with a specific wavelength triggers movement in the bonds between particles in the sample being studied, and a graph of their movements allows us to identify the material. The arrangement of the maximum point on the spectrum, that is, the spikes on the graph, is as unique as a fingerprint, although it isn't quite as straightforward as that. A simple polyethylene bottle top will generate different graphs depending on where on its surface the laser is focused. It's partly to do with additives – perhaps dyes or plasticisers – but also the structure of the object.

Scientists are worried by another matter, too. It's been established that plastics accumulate as they float on the sea, they attract hydrophobic chemical compounds known as POPs (persistent organic pollutants), which are harmful to living organisms. Many of them are carcinogens, such as PCBs (polychlorinated biphenyls), which have not been produced for a long time but are extremely persistent substances used to cool and isolate electrical systems, or PAHs (polycyclic aromatic hydrocarbons), used in the production of medicines and paint, and also a component of cigarette smoke. Plastic debris attracts pesticides, like DDT, which is banned across Europe, and bisphenol A, which has a structure similar to human hormones and so influences the endocrine system.

In some ways, it's perhaps a good thing that plastic absorbs toxic substances floating around the seas. Then again, as they break down, plastics emit the toxic monomers of which they are composed. Floating fragments of plastic shed plasticisers which make them flexible, antioxidants which delay the ageing of rubber among other things, and substances that reduce flammability or increase resistance to ultraviolet light.

The way plastics break down is so poorly understood that these synthetic materials are aged artificially in controlled laboratories – for though we know how to produce hundreds of millions of tonnes of plastic each year, we've no idea what happens to them next. To find out, researchers subject the plastics to accelerated degradation through the movement of water, sea salt, sand friction and ultraviolet light. To this day, no manufactured plastic has broken down into carbon dioxide and water in natural conditions (apart from compostable ones).

*

The term "microplastic" was coined in 2004 by Professor Richard C. Thompson of the University of Plymouth. Initially, he used it to name small pieces of anthropogenic pollution in the sediments and depths of the oceans.[55] Over the following years, interest in the issue grew and the term was no longer deemed sufficiently precise. Nowadays, pieces of plastic are split into five categories depending on their size: megaplastics, which are more than one hundred millimetres in length, macroplastics – above twenty millimetres, mezoplastics – between five and twenty millimetres, microplastics – below five millimetres, and nanoplastics – below one hundred nanometres. But let's not get bogged down in such headache-inducing details.

In Spain, "microplastic" was selected as the word of the year in 2018. It's an attention-grabbing topic, one the media covers from every possible angle. "Microplastics found in human excrement – in Poles too," cry the apocalyptic though mildly amusing headlines. Microplastics have been found in salt, honey and groundwater. They enter the aquatic environment in two ways: primarily as man-made microscopic particles used in the production of various detergents, cleansers, shower gels, creams, toothpastes and scrubs; and also through the breakdown of larger plastics. Microplastics are even released when clothes, such as those made of polyester, are washed. Fleece jackets basically scatter polyester with every step. A few years ago, it was estimated that the United States deposits one hundred tonnes of these particles into the oceans every year.[56] That number is certainly higher today.

If we're afraid of microplastics right now, we should soon be seized by nano-hysteria. Yet nanoparticles are a branch of industry that's in full-on expansion mode. They're used in paints, glues, protective layers, electronic devices and even deodorants. Exposure to heat in the course of their life cycles makes them break off from those products. Particles have been found not only in marine organisms but also in their freshwater cousins.

Don't be deceived by the nano aspect. The smaller the particle, the more easily it moves from place to place and passes through an organism's lipid structures into muscles or organs. At nano dimensions, even compounds with low toxicity can have an inflammatory effect. They influence enzyme production, growth and fertility. They impair an organism's ability to cleanse itself. Scientists have found that in the presence of polystyrene nanoparticles, mussels reduce the intake of nutrients from their surroundings.

Plastic nanoparticles move along the food chain from the smallest polychaetes and benthic organisms to huge whales, and from algae, via planktonic crustaceans, to the digestive systems of carp. Certain theories hold that the particles may disrupt chemical communication between organisms and thereby hinder reproduction, nutrition and even their ability to escape predators.[57]

The problem of rubbish in the oceans is not a recent discovery. Plastic was already being found in marine organisms in the 1960s. It's estimated that several hundred species either eat it or absorb it through their airways. Some, like crustaceans, have also learned to make use of plastic waste. There are pictures on the internet of hermit crabs using bottle tops discarded on beaches as shells for their delicate abdomens. In times before plastic, these crustaceans sought the shells of dead molluscs. Every now and then, they need to find a new home for their growing bodies. They used to fight over the most attractive shells, but plastic caps offer protection in various sizes and are often a perfect fit.

And it isn't just bottle tops that can be used as shells. Have you ever seen the photo of a crab hiding in the head of a child's doll, as if it's been watching too many human horror films? More plastic keeps coming: as it ages in seawater the rubbish acquires an increasingly porous structure and the microscopic hollows on its surface make good sites for bacteria to settle.

Floating, endlessly roaming rubbish makes for a safe and attractive haven.

*

I'm not trying, however, to give the impression that plastic in the sea can be somehow justified, or to look on the bright side of the situation. We have endless evidence of plastics' profound influence on marine organisms. Jettisoned nets, fishing gear and crab traps never stop fishing. To see the scale of the problem, it's worth going to Heligoland, a small island in the North Sea, around forty-five kilometres from the German coast. Thousands of birds nest on its red cliffs. Even with a mobile phone, it's easy to take a picture of a gannet breeding colony, just half a metre from the railings, right on the edge of the precipice. Courting begins in spring. Fortunately, the gannets aren't really bothered by our presence. The pairs flex in front of one another, delicately tap beaks and stand still, their necks embracing. What is this if not a familiar search for closeness and warmth?

I like their majestic soaring and spectacular nosedives, when they strike the surface of the water at full speed in pursuit of shoals of fish. I like those grumpy, gaggling voices, and even their rather human propensity for neighbourhood disputes. The birds belong to the Morus family, from the Greek *moros*, meaning "foolish", a stupid, unfair legacy from times when they were mercilessly murdered in their nests. Gannets only have one chick each season and they care for it very attentively. They rarely leave them on their own. But, unfortunately for them, they're not particularly scared of humans. In April, when the birds are already paired, they gather material for the nest. For thousands for years, they've used seaweed, and for a few decades now they've also used blue, yellow and orange rope or string from fishing nets.

Seaweed disappears after a season: the string will stay for ever. And here and there you can see clusters of dead, entangled birds hanging from the cliff. The bodies of gannets drying in the sun, as well as guillemots, razorbills or black-legged kittiwakes. Some of the birds that live here have no legs, they support themselves on stumps amputated by cords. Others, ensnared in fishing line, await a slow death from starvation. A scientific article from 2011 showed that each of the gannet nests analysed contained on average half a kilo of plastic. The mass of plastic in the entire colony was 18.5 tonnes.[58] Every year, sixty gannets were found trapped in string. The problem may affect eighty species of sea bird.

*

There are hundreds of photos attesting to the problem of birds caught in fishing gear, carrier bags or six-pack beer-can holders. We regularly read about large marine mammals whose stomachs contain huge quantities of human rubbish. Masses of indigestible waste clog their digestive tracts. A sperm whale that washed up on an Indonesian beach was found to contain a thousand carefully counted pieces of plastic (that were visible to the naked eye). In total, the rubbish in its entrails weighed six kilograms, most of which was made up of single-use cups and plastic bags, though it also included flipflops. There are two hundred and sixty million people in Indonesia, and 40 per cent of their rubbish ends up in the ocean every year.[59]

When animals swallow plastic rubbish, it stimulates a false sense of satiety. The stomach is always full because it can never digest a disposable lighter. The issue was brought to public attention through Chris Jordan's photographs of decomposing albatrosses with innards full of plastic: lighters, bottle tops, a whole heap of quite large, sharp fragments. A few years after taking the photos, Chris Jordan returned to the Midway Atoll to film a documentary about the lives of these majestic birds. He doesn't try to shock us with gruesome details and doesn't distract from the images with endless commentary or over-whelm us with data. The film is a homage to nature. It depicts its protagonists with empathy and love, with admiration. It tries to make us see the world through their eyes.

The birds don't feel threatened by the presence of the crew, they let themselves be observed from up close. We see their mating dance, the subtle affection they have for one another. God forbid that I should commit the sin of anthropomorphism, but I swear you can see something of ourselves in the familiar gestures of authentic bonds that unite these pairs. Midway is a paradise, although it has also been hell. In 1942, the American military base here was attacked by the Japanese. The three-day battle was won by the United States' navy and air force. The now-deserted site seems to be a place where nothing interrupts nature's rhythm. But appearances are deceptive. The ocean beaches are shockingly polluted. When the chicks hatch from their eggs, the director forces us to watch their slow agony. Many birds can't rid themselves of the toxic plastic their parents feed them. The albatross trusts the sea, the narrator tells us, each molecule of its body is built from the ocean. How can it know that the ocean is a great polymer soup?

Another indicator organism – that is, a creature that can be used to measure the degree of marine pollution – is the fulmar. At first glance, this bird closely resembles the seagull, with whom it often nests in colonies on steep cliffs. The fulmar is an exceptional aviator, seizing the slightest gust of northern wind under its narrow grey wings. It can glide for hours, just like its larger cousin, the albatross. It has a white head and a thickset neck, and its grey wings protrude from the middle of its body rather than right behind the head like a seagull's. Perhaps that's why it resembles a small aircraft adapted for flight just above the waves. But its most fascinating feature is its beak – specifically its nostrils, which take the form of a single tube, through which it expels excess sea salt from its body. It has an excellent sense of smell, which is very rare in birds, but this doesn't protect it from plastic. In the past, plastic fragments were found in less than 10 per cent of fulmars living in the North Sea. Now nearly 60 per cent[60] of the population have plastic in their stomachs.

*

What really fires the imagination? Not painstaking research, or complex, detailed analysis, but generalisations, emotive slogans based on risky extrapolations to which any honest scientist would have to append numerous unattractive footnotes. Such as this claim which made the headlines a few years ago: "In thirty years, the mass of fish in the seas will be lower than the mass of plastic." According to estimates from 2015, at least eight million tonnes of plastic end up in the ocean[61] every year. Microplastics are everywhere. Marine currents carry them to the most isolated places on the planet.

We all know this on some level, because the sea returns our waste to us every day. We recognise it as we walk along the beach, because it's universal, well-known rubbish that we all produce: bottle tops, cotton-bud sticks, plastic packaging. It's almost as exciting as picking mushrooms, especially since you come across something new nearly every day. Every walk is a surprise.

Litter-picks on beaches are an increasingly popular response to this accumulation of rubbish. On Instagram, the #beachcomber hashtag has been used almost three hundred and eighty thousand times, and #beachcleanup, two hundred and eighty thousand times. I follow a number of profiles dedicated to this noble pursuit. Like @minibeachcleaner, a thirteen-year-old from Newly in Cornwall, who shares his finds almost every day. In recent months, he's found small, plastic swimming-pool filters on an almost daily basis – an unidentified ship lost an unidentified container.

The urge to tidy up our mess is gaining ever more ground. Not long ago there was talk of plogging, which combines jogging with litter picking. Runners are meant to collect the waste they pass on their routes. "Trash challenges", spontaneous civic litter picks, are also popular, rounded off with the obligatory selfie. And I don't say that spitefully. Someone who spends four hours collecting broken bottles, picking rusty scrap out of rivers or pulling a disintegrating carpet out of the undergrowth deserves a medal. Even if it comes in the form of "likes". In Warsaw, one very active campaigner is Grzegorz, the man behind the website "Ej, nie śmieć" [Hey, don't litter], which includes lots of articles about interesting environmental news. Anna Jaklewicz's social initiative "A book for a bag of rubbish" has nationwide reach; participants receive books that have been donated by publishers.

There are also movements emerging on a larger scale. A few years ago, the "Fishing for Litter" project was launched, which pays fishermen for the rubbish they catch at sea. The objective is to clean the surrounding waters, but also to raise environmental awareness. Thirty ports and nearly four hundred vessels, all equipped with special bags for collecting rubbish, participate in the programme. In a way, it's enlightened self-interest: a KIMO report[62] states that the annual costs of repairing an engine damaged by marine litter, replacing lost fishing gear and cleaning nets is over fifteen thousand pounds per boat.

Ellipsis (formerly Plastic Tide) creates a map of pollution across the world using drones and artificial intelligence. The algorithm automatically identifies and counts rubbish in photographs, saving time and effort. According to the organisation's data, barely 1 per cent of marine rubbish is visible – the rest sinks to the seabed, is swallowed by animals or buried on the shore. In reality, we just don't know what happens to the millions of tonnes of waste that end up in the sea each year. Without that knowledge there is no way for us to effectively prevent pollution. The primary goal is to identify places where rubbish accumulates and to study trends and changes over time.

The Ocean Cleanup, a foundation established by Boyan Slat when he was nineteen, has been active since 2013. Its studies are financed by private donors as well as giant corporations such as Maersk, Deloitte, or by the Dutch government. The project, which has gone through many preliminary stages, has now moved to the implementation phase – a floating barrier passively removes plastic from the oceans, using the wind, currents and waves to move across the water. Some experts point out, however, that the device will catch floating marine organisms at the same time.[63] Slat apparently didn't know they existed. The plastic they gather is to be recycled – no mean feat given its contaminated state and the high degree of degradation. In late October 2019, the organisation presented another initiative – a device that will remove plastic from rivers. The foundation will focus on those waterways that expel the most pollution into the sea. Despite some doubts, there is widespread enthusiasm for the project – Thailand and Los Angeles County have already expressed an interest in taking part.

It seems that the constant presence of this issue in the media is leading to increased awareness. A report by the supermarket chain Waitrose contains some interesting and uplifting data – apparently 88 per cent of viewers of David Attenborough's BBC series *Blue Planet II* changed their consumer habits. After the programmes were broadcast, search engines registered a jump in the number of searches in the United Kingdom for topics related to recycling. Every year, more people take part in the Great British Beach Clean. This movement is becoming hard to ignore. The pressure forces major polluters to make commitments on reducing waste. One example is the UK Plastics Pact, a kind of intercompany alliance (including Coca-Cola, Lidl, Tesco and Unilever), whose members commit to 100 per cent of their packaging being reusable, recyclable or compostable by 2025. How touching. But what if polluters had acted twenty or thirty years ago? When environmental protection wasn't so fashionable nor such an effective tool for self-promotion.

In the 1950s, big businesses like Coca-Cola and Phillip Morris created a non-profit organisation called Keep America Beautiful, which sought to educate consumers. In the 1970s, its work was advertised by a film depicting a man in Native American dress crying at the sight of the rubbish strewn around him. "People start pollution. People can stop it," concludes the voiceover. The initiative brought about notable reductions in pollution, while also cleverly drawing attention away from the companies that produced the packaging, shifting the blame onto consumers. At the same time, these multinationals would lobby against any provisions that might require manufacturers to take greater financial responsibility or force them to design recyclable packaging.

It's worth repeating: the problem isn't the inability to recycle but the scale of plastic production, which far exceeds treatment facilities' capacity. No-one is able to process such a mountain of waste. Coca-Cola currently produces eighty-eight billion plastic bottles a year.[64]

*

I'm very fond of the apocrypha in Zbigniew Herbert's *Still Life with a Bridle*.[65] In particular, I like the letter allegedly found in 1924 in an antique shop in Leiden: three sheets of legible handwriting on well-preserved paper. They were glued into the front cover of the romance *The Knight with a Swan* in Amsterdam in 1651. The sender is Vermeer, the Dutch painter, master of silence and concentration, while the addressee is Leeuwenhoek, the scientist who perfected the microscope. It's a shame the letter is apocryphal; it would have been a real treat for historians, a look behind the scenes of two great lives, a sneak peek into the secrets of the past. But what does it matter that it's literary fantasy, since it speaks so gracefully, so beautifully and intelligently about things that really matter?

"A few days ago, you showed me a drop of water under your new microscope. I always thought it was pure like glass, while in reality strange creatures swirl in it like Bosch's transparent hell," Vermeer is alleged to have written. He describes how Leeuwenhoek reacted with spiteful satisfaction: "Such is water." Vermeer understood the allusion, he realised the scientist was expressing his superiority over artists, who record "appearances, the life of shadows and the deceptive surface of the world", and do not have "the courage or ability to reach the essence of things". Scientists as masters of the truth, artists as "craftsmen

[. . .] who work in illusion". Vermeer was riled. "With each new discovery a new abyss opens," and "we are more and more lonely in the mysterious void of the universe," he wrote accusingly to Leeuwenhoek.

Why am I giving this graceless, barebones account of a brilliant apocrypha? I had a similar experience to Vermeer, although I think I'm slightly more of a believer in science and I try not to turn away from the "new abyss". I wanted to find out what nearly broke my tooth in La Cambusa. I took my little ball to Agnieszka Dąbrowska, and she put it on a Raman spectrometer. We studied it at different magnifications: when enlarged to ten times its size, the surface had the regular ribbed texture of corduroy trousers. At fifty times, we'd gone inside the ravine of a single slit. It made my head spin – I was seeing monsters in a drop of water. A structure I had never suspected existed, a shiny surface like a lake surrounded by plastic cliffs. When magnified fifty times, the fragment of the ball looked like something from the pictures taken by drones hunting the Taliban in the mountains of Afghanistan.

Agnieszka kept altering the laser's settings: the range of light, the exposure time, the number of repeats. The computer churned out new results that brought us no closer to identifying the material at hand. The surface of the ball, worn away by waves, sand and sun, produced an upturned, arched reading. In 2018 we produced 359 million tonnes of different plastics (eleven million more than in 2017), but we aren't even able to identify them.[66] If we don't know what we're dealing with, what are we supposed to do with it next? Damaged, unidentified plastic can't be treated or recycled. We can only incinerate it. After an hour, the graph showed a characteristic arrangement of steps. Agnieszka said I was lucky – sometimes she spends weeks sitting over a single piece of plastic. My ball is most likely to be polystyrene (though there's still a degree of uncertainty).

We're more familiar with it in its foamed form used for takeaway-food cartons. Another incarnation is Styrofoam, a basic isolating material used in homes and a big problem for recyclers. It's cheap, light and highly voluminous – a huge 210-litre refrigerator would barely accommodate six kilograms of the material. No-one really knows what to do with it, especially as it easily breaks into small pieces. Transporting it isn't cost-effective. There was a glimmer of hope in the discovery that mealworm larvae eat it.[67] But it turned out that they aren't very keen to do so (they lose weight) and scientists are still unable to isolate the digestive enzyme that breaks it down. It's not really feasible to expect treatment sites to put millions of larvae to work. Polystyrene is also used for single-use cutlery (which should disappear on July 3, 2021 thanks to an EU directive), toothbrushes, shavers – that is, a whole load of the cheap, unnecessary junk. I don't know why I thought my ball might be something exceptional, something rare or precious. Instead of poetry, I got prose – as plain as polystyrene.

AUGUST 8, 2018

c.2.15 p.m., 50°40'33.8"N, 22°53'33.4"E

The forest was silent. Empty bottles of chainsaw oil and beer cans were scattered all over the place. I was patiently picking them up when I spotted a cylindrical silvery aerosol poking out of the mud. On it, I saw a picture of a fly and the text: дихлофос препарат инсектицид. I read the back of the dented can. Price: one rouble. Manufacturer: Slavgorod PO Altaikhimprom. Year of manufacture: 1991. The final stages of the system's delirium. A growing budget deficit, rampant crime, a month's salary buys you two packs of Marlboros.

But though the end of the world – sorry, the Soviet Union – is nigh, far away on the plains of the Kulunda Steppe, the Altaikhimprom factories keep working. On one of the plant's buildings, a sign reads: "To live better, you must work better".[68] In 1991 that might sound a bit ironic, since however hard they work in Slavgorod, four hundred kilometres from Novosibirsk and the same from Omsk, there is nothing there to herald a better life. Altaikhimprom still exists today, but in the intervening years it's faced a number of financial scandals. At the time of writing, the plant's webpage redirected to a porn site.

And the substance itself? Dichlorvos (as it's known in English) was intended as an insecticide, but was always somewhat suspect because it did rather more than simply kill flies. It had quite diabolical ancestors – nerve agents used in chemical warfare – and dichlorvos didn't entirely lose those predatory instincts itself. On the International Labour Organisation's website, the description of the substance is dotted with exclamation marks. Dichlorvos is potentially lethal if swallowed or in case of prolonged contact with skin, and long-term exposure can damage the nervous system. It is highly toxic to aquatic organisms. Its use is banned in the European Union and the UK.

What was dichlorvos doing in the middle of a beech forest in the heart of the Szczebrzeszyn Landscape Park? Altaikhimprom

products can be obtained across the whole territory of the former USSR. The dichlorvos must have found its way onto market stalls in nearby Zamość, on camp beds in Stefanides Square or in the city park. This can was then bought by someone who didn't appreciate being bitten by insects. They used dichlorvos, no doubt completely unaware that it could damage their own DNA. Perhaps they sprayed it on their children, not knowing it can lead to ADHD.[69] The dichlorvos ran out and the can ended up on the slope of the gorge. How many years had it been there? The writing is still legible. Aluminium resists corrosion because of passivation – it's gradually covered in a watertight layer of oxides that acts as insulation.

Dichlorvos can't legally be purchased in Poland, but there's no such problem in Ukraine. It comes in all sorts of varieties. When you order it wholesale, it works out very cheap, just twenty hryvnia (around fifty pence) per can. For that price, you can buy the version that bears the compelling name, Morfei (Morpheus). If that's too literal, there's also Dichlorvos Eko, with a lavender scent. On the consumer review website otzovik.ru, the user Rusalochka from Omsk writes that she's used dichlorvos since she was a child. Her father was in the army, which meant they frequently moved house, and every time this brilliant insecticide came to their aid. I picture those grim Khrushchyovka apartment blocks in places like Khandyga or Srednekolymsk, with cockroaches the size of Moskvich cars. Rusalochka welcomes the fact that dichlorvos now smells like lavender, because that suffocating, chemical odour was unpleasant. The price is still good. Rusalochka hopes dichlorvos will be around for a long time to come.

4

LEONIA

If you were writing a synopsis for a TV listing, it might be enough to say: "Marco Polo tells the elderly Kublai Khan about the cities he has visited". But summarising the fleeting action makes no sense because that isn't what matters in Italo Calvino's *Invisible Cities*. Whenever I pick it up, I lose myself in this novel from 1972, which is also an elaborate treatise on storytelling and the imagination. Some episodes are only loosely tied to our reality, whereas others present an overwhelming dystopian vision that sounds worryingly familiar. Like the story of Leonia:

The city of Leonia refashions itself every day: every morning the people wake between fresh sheets, wash with just-unwrapped cakes of soap, wear brand-new clothing, take from the latest model refrigerator still unopened tins, listening to the last-minute jingles from the most up-to-date radio. On the sidewalks, encased in spotless plastic bags, the remains of yesterday's Leonia await the garbage truck. Not only squeezed tubes of toothpaste, blown-out light bulbs, newspapers, containers, wrappings, but also boilers, encyclopedias, pianos, porcelain dinner services. It is not so much by the things that each day are manufactured, sold, bought that you can measure Leonia's opulence, but rather by the things that each day are thrown out to make room for the new. So you begin to wonder if Leonia's true passion is really, as they say, the enjoyment of new and different things, and not, instead the joy of expelling, discarding, cleansing itself of a recurrent impurity.

*

In our world, and especially in cities, many things occur unnoticed, beyond the limits of our awareness. We pay for convenience. We pay for electricity and gas, we contribute to our apartment block's maintenance fund so the roof doesn't leak. Someone takes care of it; someone makes sure there's running water in the tap. It's no different with our refuse: someone makes it disappear. In Poland, nearly 12.5 million tonnes of municipal waste were collected in 2018[70] (13.5 million according to industry data). Our daily rubbish. That number is rising, slowly but surely. At the moment, it works out at 325 kilograms per person per year, which is still modest in comparison with citizens of the West. The EU average is 489 kilograms.[71]

*

I wanted to find out what happens to my carrot peelings, my cream-cheese cartons and my cat litter, but the trail went cold surprisingly fast. Warsaw's Municipal Waste Company (MPO), which collects waste from my house, kindly agreed to help me. But it turns out that "collects" doesn't mean "treats". The MPO passes my rubbish on to another company. Dealing with them wasn't as straightforward. I wrote, I called, I listened to assurances that someone would get back to me. But "someone" decided that my visit wouldn't serve the company's interests, and I wasn't determined enough to creep around a treatment facility in the dead of night. If it's so difficult to trace our own innocent rubbish, what hope is there for the hazardous waste shamefully concealed in developing countries?

Speaking of which, I often hear that while we enlightened Europeans are conscientiously separating our rubbish, somewhere in Asia people are throwing all theirs straight into the sea. In essence, the stereotypical natives of those poor countries we fly to for cheap holidays sabotage our hard work and our sacrifices in the fight for a clean environment. People are genuinely indignant – after all, they take reusable bottles to work and pack their shopping in bags made of old net curtains. And yet few things seem so unwarranted and unfair as blaming countries of the so-called third world for polluting the seas. We pay no attention to the fact that our disappearing garbage is one of the privileges of living in the developed world. In countries where the majority of people live below the poverty line, there is no infrastructure for the collection and management of waste.

Today, we also know that the well-organised, clean Western world sends masses of its waste to places where no detailed regulations exist. To places inquisitive activists or reporters will not find it. To the other side of the world, so far away as to leave us basically indifferent. In fact, a lot of it also comes to Poland. I read that one party pledged in its manifesto to close the borders to foreign waste. That's probably good for us, but the rubbish won't just vanish. There's sure to be a country willing to take it. Why do we behave as if we're living on different planets?

*

My ride on a bin lorry began with a language lesson. Or more accurately, a lesson in empathy. Words matter. So let's not say "bin man". It's not a question of having a particularly fine ear or being over-sensitive, anyone can hear the contempt hidden in that word. I'm not surprised the people who clean our streets dislike it. In the language of industry, they are refuse collectors. That may not roll off the tongue, but tough: we'll have to force it.

And what about bin lorries? That's a word over which no-one will quarrel, it's more practical, it doesn't offend anyone. But in formal language, a bin lorry isn't a bin lorry. It's a refuse-collection vehicle.

The work of a refuse collector is not respected. They deal with waste, after all, something unwanted, thrown out, spurned, something that is to be removed from our sight. Our attitude towards it colours the way we treat the people who collect it from our homes. And this becomes all too clear as the lorry stops at each address and a small traffic jam forms behind it. Any delay and cars start honking their horns. Angry people step out, shouting that they have to get to work, to go about their business. They make impatient gestures, start making phone calls. They don't have time. "Hey, how much longer will this take?"

For a while I daydream about a great rebellion of collectors and drivers who say, "To hell with it," and let the city drown in rubbish. All those busy people suddenly forced to fight their way on foot down corridors of waste. Buried in avalanches of kitchen paper, watermelon skins and meat packaging. Their brows sweaty, they try to scale the mountains of garbage to work out where they are. They cut themselves on broken glass. They tremble in fear of meeting huge, emboldened rats. I think about how the city would smell, how we'd learn to live with the stench. The privileged would no doubt have special odour absorbers. Expensive, energy-intensive ones. And special enclaves surrounded by a high wall, so they and their children could live in a clean environment.

The drivers sound their horns at the bin lorry. No-one would dare honk at a fire engine. No-one would yell at an ambulance driver. When we reach a gated community in a new, pretentious neighbourhood, the security guard follows our every step, though his lack of interest is obvious. He's just pretending to be a vigilant guardian. Apparently, the residents don't like strangers hanging around their grounds. The refuse collectors meanwhile go about their work in silence, without resentment. That's just the job. You don't work in waste collection to follow a vocation or for pleasure. It's never been a prestigious occupation, though in the past it was apparently well paid.

In the early 1990s, one collector, Marek, worked as a driver for the Polish cabinet office, ferrying bigwigs back and forth, but was barely able to make ends meet. He found that he could make twice as much in the waste sector. In the past, collectors might earn a second salary on the side – people always have something to shift, after all. Marek has been working on the bin lorries for nearly thirty years now and he's seen a few things. And heard them too, but at that he laughs cryptically. On the new estate, rubbish is separated into different fractions depending on its type, but in the black bin for mixed waste, we see everything: food leftovers, nappies, boxes from meal delivery kits, bottles, rubble from minor home repairs. The container for plastics holds a single stiletto shoe.

The bin lorry is new and very clean inside. Tadek, the driver, gives me an amused look and suggests I spend an eight-hour shift in it if I want to notice the shortcomings. "OK," he admits with a drawl, "the new one looks nice, but the old, ugly one, well, it was roomy!" It's true – for three men this one is a squeeze. The person in the middle has to tuck their legs in and sit sideways on, and they spend most of the day in that position. And it's hard work, after all, and you have to get on with it whatever the weather. In November it's pouring with rain, in February the cold numbs your fingers, and in July you face searing heat and a stench so bad it makes your head ache. But you can get used to the smell: it's more difficult when you come across residents who treat refuse collectors like servants, someone to take their frustrations out on, someone they can lecture or humiliate.

*

The bin lorry makes me think of a large, round-bellied insect filling its stomach. Everything related to the waste industry brings to mind weird analogies involving the digestive system. This endless cycle of consumption and excretion. The iron claws lift the bin and pour the contents over the back of the vehicle, straight into the compactor. From there, the rubbish falls into the drum. The onboard computer receives a signal from a chip on the bin, and a panel on one side of the lorry displays the address, the mass of the load and how many collected bins have been paid for. The information goes to a database. Our waste is scrupulously counted, and the enduring tale of the bin lorry that "chucks all the rubbish together" is an urban myth. Today, each fraction is collected separately (sometimes trucks have different compartments for different types of waste).

But, before the bin lorry makes an appearance, the journey of an item that has reached the end of its useful life and been turned inexorably into waste, begins in the household dustbin. Or rather in the bin liner. The bin liner was a dazzling child of the 1950s, the decade that saw the start of the plastic boom and the birth of "throwaway culture", and which revolved around single-use items meant to be discarded forthwith. The liner is so hygienic and convenient that its place in the home now seems unquestionable. Who isn't repulsed by the ever-filthy organic waste caddy or the mixed black bin, after all? The bin liner was invented by the Canadians Harry Wasylyk and Larry Hansen. Wasylyk created a low-density polyethylene prototype in his kitchen using a method known as extrusion, that is, by squeezing heated plastic into a mould of the desired shape. The earliest bags were green in colour and were produced for a Winnipeg hospital to stop the spread of the polio virus. Mass production began in the late 1960s.

*

Another shift on the bin lorry. This time we're collecting waste from my old neighbourhood, including from the apartment block I lived in for over thirty years. I haven't been here for some time, but it doesn't look like much has changed. I expect any waste segregation to be mostly symbolic, because the twelve-storey building has a rubbish chute, a relic of a bygone era when everything went to landfill. There are bins for separated waste, but they're several hundred metres from the front door and, judging by what we find inside them, not many people take the trouble.

So this is the mixed waste fraction, on the first collection after Christmas: stacks of wrapping paper, gift bags, greetings cards, ribbons, cardboard packaging from toys. Piles of packing slips, invoices, orders for presents, cardigans with pockets and sweatshirts with a bulldog on the front. Letters to Father Christmas and old school tests. Dorotka's A-grade maths test. Question 4: "Weronika had 20 złoty. She bought a set of cards for 8 złoty. How much change did she receive?". Answer: "Weroinka recieved 12 złoty." Dorotka holds more promise as a mathematician. Aside from all that, the containers are full of tissues, which isn't surprising given that the temperature is hovering around zero and it's the annual flu season. And there are a lot of paper towels – so convenient, disposable and hygienic. In the United States, 3.3 billion kilograms of used toilet roll, kitchen paper or tissues are thrown away every year,[72] with no hope of being recycled. I think of the trees, how the last traces of their decades-long lives are vanishing. Just as a fillet barely resembles a chicken, a tissue doesn't resemble a tree.

Another lesson on board the bin lorry: an introduction to the secret life of the streets of my childhood. At number 21, Marek duly warns me: "Careful, this place always has cockroaches." And right on cue, we find the insects crawling in their repulsive way all over the bags. At the next stop, Marek tells me: "It's going to stink." Oh God . . . I gulp even as I write these words because the stench left me bent double. I couldn't step inside the shed. In the bins at my primary school, the canteen slops had been rotting in plastic bags for a week. At the kebab stand by the shops, a rat scurries from beneath the container. That's nothing to do with Turkish cuisine – apparently you often come across large, bin-dwelling game outside all bars and restaurants.

OK, so what needs to be improved? Refuse collectors have no doubt that plastic is not collected frequently enough – once a week in this neighbourhood. In other places, bins are emptied more often, even though the regulations (set by Warsaw City Council) state that collections should be made at least once per fortnight. I struggle to imagine what my bin would look like if left unemptied for two weeks. When the containers for plastics are full, people discard their carefully segregated waste into the mixed bin. At the apartment block where my maths tutor used to live, someone has thrown some smashed baubles, tangled tinsel and a gordian knot of Christmas lights into the bottle bank. Can a bauble be recycled? I don't know, but there's no point deluding ourselves with a fantasy of endless recycling, of materials circulating in a closed loop forever. Perpetual motion is impossible.

*

The fact is that street cleaners are welcomed like angels, and their task of removing the residue of yesterday's existence is surrounded by a respectful silence, like a ritual that inspires devotion, perhaps only because once things have been cast off nobody wants to have to think about them further.

Nobody wonders where, each day, they carry their load of refuse. Outside the city, surely; but each year the city expands, and the street cleaners have to fall farther back. [...] Besides, the more Leonia's talent for making new materials excels, the more the rubbish improves in quality, resists time, the elements, fermentations, combustions. A fortress of indestructible leftovers surrounds Leonia, dominating it on every side, like a chain of mountains.

[...]

The greater its height grows, the more the danger of a landslide looms: a tin can, an old tyre, an unravelled wine-flask, if it rolls towards Leonia, is enough to bring with it an avalanche of unmated shoes, calendars of bygone years, withered flowers, submerging the city in its own past, which it had tried in vain to reject, mingling with the past of the neighbouring cities, finally clean. A cataclysm will [... cancel] every trace of the metropolis always dressed in new clothes.

*

BYŚ is a company that collects and treats waste from a few Warsaw neighbourhoods and suburban municipalities, and it has kindly agreed to my request to take a closer look at the work of a treatment facility. Their plant is modern, so they have something to show off. Up to three hundred thousand tonnes of municipal waste can be sorted here every year, about half of which is mixed. I quickly get lost in the maze of conveyor belts that look like they could have been designed by Escher. Some move slowly, others more quickly, and after a moment I'm no longer sure whether they're travelling upwards or downwards. First, the mixed fraction is separated into two groups inside a drum. Small pieces of metal and plastic are recovered, then organic waste measuring up to eight centimetres in size is stabilised (through aeration and hydration in order to prevent the waste from fermenting and to make it as harmless as possible to the environment). This is known as the undersize fraction. The larger, inorganic waste, known as the oversize fraction, is sent down a complicated treatment line.

It starts with magnets, which remove ferrous and non-ferrous metals from the stream, followed by optical sorters, which deal with the plastics and paper. The belt races along so fast I can't keep up with single pieces. How does the sorter work? The device sends out infrared beams to identify the material, and then a carefully calibrated nozzle blows the items into the right container. Any waste left unsorted travels back down the belt again.

There are also people at work in the sorting plant; beavering away in little cabins, they remove cardboard, clear and coloured plastic wrap, metals, and bottles made of PET and glass from the belts. Someone has to correct the work of the machines. As we walk past, *perestayut rozmovlyaty* (Ukrainian for "they stop talking"); the quiet, calm, hardworking agents of our economic growth. It's a little smelly in the hall, but much less than I expected. A spray cannon fills the air with a mist that neutralises the odour of the rubbish. BYŚ also has one of Poland's few installations for the treatment of construction waste. Rubble is turned into aggregate, which can be used, for instance, as ballast for roads.

Disposal comes at the bottom of the waste hierarchy outlined in EU legislation.[73] It's supposed to be a last resort. By law, all rubbish, including mixed waste, should be sorted before going to landfill. Since 2016, it has been prohibited to dump waste with a calorific value of more than six megajoules. This covers plastics used in food packaging, for example. But is that what happens in practice? Unfortunately, in Europe (and Turkey) only 32 per cent of all collected plastic waste is recycled, 25 per cent is landfilled and 43 per cent incinerated[74] (in these statistics, 4 per cent vanishes without trace).

How is it possible that despite these regulations, plastic still doesn't end up where it should? In Poland, the law leaves some margin for creative interpretation. Many facilities that collect waste assume consumers have sorted it at home. Then the mixed waste goes straight to the incinerators or to landfill. To properly sort and treat it, you need a modern site. Not all can afford the right equipment. Not all facilities are equal – the sorting systems can be meticulous and involve multiple stages, or they can be entirely symbolic. The sad truth is that the fate of our rubbish varies enormously depending on where we live and on the company that won the tender for refuse collection.

*

Nowadays, everything is counted, inventoried and entered into Excel. You can plot a graph. The data always shows something. Numbers are cold, precise and perfectly emotionless. Objective. Our whole world is based on this assumption, even though it's false. How do you reach recycling levels? Through minor manipulations. The West disposes of its waste by selling it to less developed countries. The UK, for instance, doesn't have the infrastructure to treat its millions of tonnes of waste, so it exports them. It's cheaper that way. Officially, packaging is sent for recycling, but in reality it ends up in landfill. The consignors aren't too bothered by the fate of their waste, they come away with a clear conscience and documents to attest that recycling took place. The statistics add up. Now, is everyone satisfied?

The problem is that waste doesn't just disappear if we close our eyes. The issue became urgent in early 2018, when China, the biggest importer of plastic waste in the world, launched a crackdown. They had too much rubbish, pollution was starting to get out of control. Now they only accept clean plastic that is actually suitable for treatment. So Western countries quickly found other markets for their refuse. In principle, there is nothing unusual about the cross-border movement of waste.

Many specialised facilities import specific types of waste from the other side of the world, so long as it pays off. But we're talking about garbage – useless, dirty waste that can at best be incinerated. Despite the fact that it is incapable of managing its own waste, Poland imports a lot from abroad, and it isn't necessarily just the recyclable material cited in official data.

Importing rubbish seems even more dubious an idea when you consider the fact that the number of fires at Polish landfill sites has been increasing in recent years, reaching a peak in the spring of 2018. It was becoming clear that the waste kept in warehouses and huge hangars wasn't suitable for treatment. When space grew short, or when permits were about to expire, mountains of rubbish were sent up in smoke. Pending a full investigation, it was suspected that this is what lay behind the week-long fire at the huge landfill site in Zgierz, which burned down at the end of May 2018. The British were forced to get their tongues around that place name when it transpired that hundreds of tonnes of their conscientiously separated waste had ended up on a garbage mountain owned by GreenTec Solutions, the company that was supposed to recycle it. In a depressing dispatch from the site of the fire, David Jones, a reporter for the *Daily Mail*, described the jars of Lloyd Grossman pasta sauce, Sainsbury's roast ham packets and many more bits of typically English rubbish.[75] Experts later estimated that the plant's infrastructure was in no way capable of processing such huge quantities of rubbish, and the quality of the materials collected there was so poor it would be impossible to recycle them properly.

And Poland is just one of many countries that decided to profit from the closure of the Chinese market. A deluge of rubbish has inundated countries like Malaysia, Indonesia and Vietnam – countries where large quantities of plastic have no hope of being properly recycled. In the first half of 2018 alone, Malaysia accepted nearly half a million tonnes of plastic from abroad. Even incineration takes place here without any kind of standards being maintained. In the end, the governments of these countries had to take radical steps to prevent an environmental disaster. Many containers of polluted municipal waste, labelled as plastic for recycling, were sent back to their countries of origin: the UK, Spain and Canada.[76] And to the United States, where the practice of sending rubbish to countries with no regulated standards for waste management dates back to at least 1992 (when the first evidence of it appeared).[77]

*

Karol Wójcik, BYŚ's spokesman, talks to me about the situation in the waste industry. There are rays of hope. 2020 saw the launch of the Waste Database, a convenient, freely accessible tool that lists all relevant stakeholders – those who put products that require separate collection on the market, and the firms that manage waste. It allows anyone to check whether companies have the correct permits and competences. Every shipment of waste is recorded. The database is supervised by the Environmental Protection Inspectorate, which enjoys extensive powers. The main problem at the moment is the lack of professional waste management installations.

Of course, there are still companies that mismanage waste and dispose of it illegally. Bertold Kittel offered an interesting insight into the waste underworld in his book, *System. Jak mafia zarabia na śmieciach* [*The System: How the Mafia Profits from Waste*].[78] Although it was published in 2013, many issues are still relevant today. Every year, Poland is shrouded in smoke from torched landfill sites, as fraudulent companies dispose of their rubbish, hazardous waste included. These businesses function at the limits of the law, often without the appropriate permits, frequently registering their activity as something else entirely. A tall fence guards the secret. Without strong grounds for an inspection, the authorities are in no hurry to intervene. In this context, tightening the rules above all affects the honest businesses that do want to comply with requirements. The dishonest ones won't change, the rules don't apply to them. But, hopefully, thanks to the Waste Database, the situation will improve.

The system's second major problem is money, or rather the lack of it. In Poland, there are rules that oblige manufacturers to finance the collection and recycling of the items and packaging they produce. In principle, at least, the polluter pays. Company X makes fizzy drinks and arranges for its bottles to be collected and duly treated so there's a chance that in some form, in some recycled reincarnation, they'll go back into circulation. X doesn't do this by itself, however. Instead it pays for the services of a recovery organisation that acts as an intermediary between X and the treatment facility. Organising the process in this way should ensure that collection is more efficient and treatment more refined.

And yet in Poland, manufacturers' payments into the system are often several dozen times lower than in the rest of Europe. In August 2019, the charge to the producer for one tonne of plastic was 0.6 euro. By way of comparison, in the Czech Republic it's 206 euro. This disparity applies to all materials. For one tonne of glass, the charges are 5.4 euro in Poland and 67.7 euro in the Czech Republic. For paper: 1.5 euro and 91.19 euro. In relative terms, the contrast with the Czech system isn't even that stark. If you were to compare the costs incurred by Polish producers with the fees they'd face in Austria, the chasm would be enormous. Manufacturers say increasing the charges would lead to significantly higher prices, but that isn't true – retail prices in Poland and the Czech Republic are very similar. Waste management as a whole would be more effective though, so perhaps the overall cost to consumers and taxpayers would decrease.

The third problem that persists, though hopefully not for much longer, is the issue of waste separation. Or to be precise, the fact that, until recently, it wasn't mandatory to sort your rubbish in Poland, and the charges for collecting mixed waste were only slightly higher than for separated waste. From September 2019, local authorities were given a year to adapt their by-laws to the new legal framework. Sorting rubbish is no longer a magnanimous choice, but an obligation. The charge for disposing of mixed waste will be two, three or even four times greater than for segregated waste. It's the only way to meet the growing demands of the European Union – by 2025, we're meant to reach a target of 55 per cent of all municipal waste being prepared for reuse or recycled, which experts say is an impossible task.[79] Without the threat of higher charges, there is no real incentive to encourage citizens to sort their waste, beyond the environmental argument. Apartment blocks pose their own problems. What does it matter if most residents dutifully separate their rubbish when there'll always be the odd black sheep who pours rancid beetroot soup into the paper bin? In cases like this, collective responsibility will apply. Everyone will pay for the carelessness of a minority.

Of course, for many of our compatriots, environmentalism is just a pain in the neck. One often hears complaints that the separation system is overly complex. Admittedly, some pieces of rubbish confound even workers in the waste industry, but there are many basic, reliable guidelines on how to sort it. Practically every local authority publishes their rules on their websites, as waste collection varies depending on where you live. In Szczecin, as in Warsaw, waste is sorted into five fractions, but for some reason you can't put egg shells in the organic bin.

I have no way of knowing what will happen to my rubbish once it leaves my bins, but I do follow a few rules: I put all plastics into the container for plastics and metals, even pieces that are too small or too difficult to recycle (such as anything labelled "7"). I separate materials – removing the aluminium lid from a polypropylene pot, for instance. Polystyrene trays, greasy paper, ceramics, tiles, wood, pieces of cloth and any irredeemably dirty items go in the mixed waste. I take clothes (that can genuinely no longer be worn) to clothing banks so there's a chance they'll be used as some kind of stuffing. Lightbulbs and small electrical items can be taken to a local recycling centre or there may be collection points for them on the high street. It really isn't that complicated.

*

Rubbish is revolting, it's disgusting . . . That is, unless it belongs to somebody famous. Anything to do with celebrities is exciting: what they wear, what they eat, what sports they enjoy. Our desires and ambitions are piqued by images of lives we can't have. Photographs of the huge estates, luxury yachts and parties where such demi-gods gather leave us feeling frustrated, so it's no wonder we also want to see what goes on behind closed doors. Any proof that they're human after all, any weaknesses that might hint at their downfall, are comforting. Perhaps that explains the tremendous appeal of Bruno Mouron and Pascal Rostain's photographs in *Paris Match*?

For fifteen years, the photographers rummaged through bins and took pictures of celebrities' rubbish, first in France (targeting Brigitte Bardot and Jean-Marie le Pen, among others) and later in the United States (where Jack Nicholson and Ronald Reagan were among their subjects). The waste told a story about the habits and consumption patterns of the rich and famous. Nicholson drank Corona beer, ate Doritos and used Downy fabric softener. It's a real litany of trivialities. Fortunately, they also exposed a few minor celebrity sins. It seemed that the Republican Arnold Schwarzenegger enjoyed Casillas cigars imported illegally from Cuba, which was still under a trade embargo at the time.[80] After a smoke, he'd freshen up with Binaca breath spray.

The left-wing agitator Alan J. Weberman acquired fleeting fame among those riffling through other people's rubbish when he went through Bob Dylan's bins. Among other things, he found the singer's song manuscripts and notes. His dig through the dustbin is shown in the documentary, *The Ballad of A.J. Weberman*.[81] An irritated Dylan once confronted his stalker in the street and also sued him for invasion of privacy, but the court ruled that rubbish outside one's home is no longer private property. Weberman wrote a concordance of Dylan's songs (an alphabetic list of phrases occurring in his work), together with his own interpretations of them. He proclaimed himself a Dylanologist and over time he even managed to become somewhat acquainted with the singer and speak with him over the phone. Recordings of their conversations can be found online.

*

The heavy metal shutter slowly rises, letting in the hazy wintry light to reveal a heap of rubbish. It doesn't even smell too bad, perhaps because of the cold. On the surface, I can make out some packaging, plastic wrapping for toilet paper with a little fox on it, yogurt pots, clothes, bin bags. In the semi-darkness, whatever lies beneath the top layer is already a compact, homogeneous mass. A metal claw, known as the grab, travels along a horizontal crane and carefully plucks some rubbish from the pile. This waste is destined for incineration and what's known as energy recovery.

The rubbish is thrown into a furnace from a height of around fifteen metres, and the flue gases are sent into the afterburner chamber. "If you, Mr Łubieński, were to set fire to some tyres, they would release black, acrid smoke. That smoke is flammable, it can be burned in an afterburner chamber and reach an even higher temperature. In our furnace, the temperature never falls below 850°C. It's now at 1080°C," explains Roman Miecznikowski, director of waste management at the Targówek incinerator on the outskirts of Warsaw. And what if the temperature dropped? "We'd pay a fine. Rubbish combusted at low temperatures emits more harmful substances." It's the job of the crane operator to maintain the required temperature – he or she uses the grab to pick up the rubbish, estimating its calorific value from its mass. The heavier the load, the damper it is, which means it won't burn so well. A batch of this kind requires additional calories, so some fragmented bulky waste (such as furniture), or the remnants of separated waste that can't be recycled, are added.

The temperature in the furnace and the levels of emissions are constantly monitored in the control room, while the boiler, a tangle of pipes and some of the filters are locked in a huge hangar. The machinery is so loud I can barely hear Roman's explanations. I have to concentrate because it's quite complicated. The flue gases from the afterburner chamber are channelled into the heat-recovery boiler, where the "recovery of thermal energy" takes place. The heat from combustion is used to heat water flowing through the boiler. The water turns into steam. Once it reaches certain conditions, it's directed into the turbine set which includes turbines and a generator. These installations are in a different room. Energy from the steam is turned into mechanical energy, and the rotation of the turbine blades drives the generator shaft, producing electricity.

In winter, some of the steam from the first stage of the process heats water that ends up in the district heating network. However, the Targówek site recovers little energy and most is spent covering the needs of the incinerator. It's quite an old installation, still awaiting modernisation. When it's upgraded, it's performance will increase significantly, as will the amount of energy recovered.

Afterburning is all very well, but combusting rubbish produces a huge amount of pollution. The flue gases represent the bulk of it. First, they're sprayed with ammonia to eliminate nitrogen oxides. Next, after passing through the boiler, they end up in the absorber, where they are cooled. Phew! Now comes filtering. The flue gases are mixed with hydrated lime, and together they are sent to the bag filters. These mostly catch dust but also chlorine and sulphur particles. After all of these steps, the flue gases end up in a huge silo in front of the hangar. That's the adsorber, where activated carbon removes dioxins and furans. The rest is released from the eighty-metre-high chimney. Roman says the smoke coming out of the chimney is cleaner than that of a cigarette.

It's often said that incinerators help to save space in landfill sites. Indeed, in the Warsaw incinerator, 150 tonnes of mixed waste are reduced to 40 tonnes of bottom ash, and in new installations, the reduction in volume can reach 90 per cent. But the remains are unstable, which means they have to be disposed of very carefully. The used activated carbon filters also head for the dump. At least the incinerator has a holding tank, which prevents any residues from being washed off the site's land. Polluted rainwater ends up in settling tanks.

There are now twelve incinerators in Poland, and more are under construction. But environmental organisations insist that burning rubbish is wasteful: you can only incinerate everything once.

*

Tom Szaky is the founder of TerraCycle, a company that recycles waste. Waste of every kind. You could call Szaky a visionary, a sorcerer, or perhaps a dreamer, because he offers a beautiful image of the world, in which absolutely everything we throw out has value. And for that reason, every item of rubbish deserves close attention. Szaky's book, *Outsmart Waste: The Modern Idea of Garbage and How to Think our Way Out of It*,[82] is surprisingly optimistic. Positive energy permeates its pages. The author doesn't torment us with stories of turtles eating plastic straws. Of course we have a problem, but it's not as if we're powerless to address it. Szaky maintains that rubbish is actually one of the few issues over which we do have some kind of control.

Rubbish is a human concept, unheard of in the wild. In the natural world, everything has its goal, its place and its purpose. A healthy system regulates itself; matter circulates endlessly in a closed loop. Excess is almost non-existent: there are numerous factors that limit access to food, available protection and the ability to reproduce. Humans have learned to remove such barriers, hence the outrageous festival of wastage that is our daily life. While millions of people starve, inconceivable quantities of food go to waste across the world every year.

Szaky argues that recycling makes sense, but it has to be worth it. Producers should pay a tax that factors in all environmental aspects of manufacturing (emissions, energy consumption and the impact of extracting raw materials on the surrounding area). This money would allow for the development of technology that would ensure even problematic waste went back into circulation. Of course, it would mean certain objects became more expensive. TerraCycle seeks to prove, as a matter of principle, that everything can be recycled, even nappies. In Amsterdam, they are collected at locations across the city, sterilised, and then separated into cellulose, plastic and the material that absorbs and binds urine. New nappies can then be produced from these components. Used cigarette filters can also be recycled. The tobacco traces and cigarette paper are composted, the filters are turned into plastics. One partner in this initiative is the city of New Orleans, where thanks to

TerraCycle's system of dedicated bins, the number of cigarette butts found on the street has been reduced by 13 per cent.

*

Every way you look at it, recycling makes sense. Our planet can't afford throwaway living. We're irreversibly depleting non-renewable resources, and the yield of renewable resources no longer meets the demands of the global market. We simply can't manage without recycling. Take paper, for instance, which is made of cellulose from wood pulp. Trees are planted just like any crop and after a few decades they're cut down, but with the current demand for paper, relying solely on the original raw material is impossible. Four hundred and thirteen million tonnes of paper are produced every year. Recycling a tonne of wastepaper brings an energy saving of 75 per cent, significant reductions in air pollution and uses a third less water than manufacturing from pulp. It also saves seventeen trees.[83] If I could choose, I'd save the monumental beeches and firs in the natural forests of the Bieszczady mountains, which are being felled without reason or mercy.

Aluminium recycling brings truly spectacular savings for the environment. Producing aluminium from its raw materials, principally bauxite, is a complicated process. First the rock must be milled, then comes digestion, precipitation and calcination, after which the aluminium oxide is electrolysed. The process is extremely energy intensive. Recycling uses only 5 per cent of the energy required to produce metal from bauxite. It also means reduced emissions of harmful substances into the atmosphere and less water pollution. It's win-win. The high price of scrap contributes to the decent levels of collection in our country – aluminium always finds its way to the recycling centre.

But of course, what we're more interested in is the recycling of plastics. The data we have reveals the sad truth that only 9 per cent of the plastics produced by mankind have been recycled.[84] Who knows, perhaps even that grim figure is overly optimistic. In Poland, recycling levels are low. The efforts of people who carefully separate the aluminium lid from the polypropylene pot are to a large extent thwarted by those who chuck everything out together – particularly in towns, where rubbish is essentially anonymous. Dirty plastics reduce the quality of the recyclate and recyclers don't want to take them. Besides, it isn't easy to sell recyclate from recovered plastic. The original resources are still very cheap, which is why barely 7 per cent of annual production uses recycled material. Large quantities of plastics that are fit for treatment are waiting in the forecourts of sorting centres for a better economic outlook, for higher prices. We can only hope that time will come before another fire conveniently removes the problem.

There's no such thing as a closed loop for plastic recycling, either. There are three problems worth mentioning. First of all, there are the plastics themselves: polymers consist of long chains of monomers with the addition of various fillers. During treatment, these chains may be shortened, and in order to obtain a high-value product, many of them require the constant addition of new plastics. Not all plastics are equal – recycling polystyrene isn't currently viable, for instance. The second problem concerns the impurities that can accumulate in plastic waste. Finally, the recycling process also creates by-products that are harmful to the environment.

And quite simply, not everything can actually be recycled. After sorting, you are left with large quantities of the under-size fraction, the smallest pieces, that can't be reused. Once stabilised – that is, once any organic traces have been cleared – it consists almost entirely of plastics. They can be used to make so-called alternative fuel, which is used in cement factories, for example, but as waste production increases, there is ever more rubbish that cannot be recycled. In Poland, it's reached the point where sorting centres have to pay cement factories to take the material, whereas just a few years ago, the opposite was true. There's just too much plastic.

To this day, no recycling installation has been built that is efficient enough to achieve satisfactory results. The technology is expensive, as is the equipment; everyone looks for short cuts, and meanwhile, plastic production continues to grow. Some in the waste industry suggest that this surplus could be used for co-generation: energy from incinerated rubbish could heat the water in our homes and produce electricity, like in Scandinavia.

Sweden, for instance, relies on incinerators for much of its electricity and thermal energy. Yet until recently government webpages stated that nearly 100 per cent of all waste was recycled. A case of minor fraud, or perhaps creative accounting, as someone more forgiving might say. Energy recovery is not recycling.[85] Environmentalists stress that incinerating well-sorted, clean plastic is wasteful. Oil, the basic component of plastics, is limited, it's non-renewable, and the market price of a barrel does not reflect its true value. A few years ago, British Petroleum estimated that at the current rate of extraction the deposits will run out in about fifty years. And meanwhile we're burning plastic packaging because it's calorific – it keeps the temperature high in furnaces.

Denmark has an even longer tradition than Sweden of burning rubbish. The first incinerators were built there in 1903, and nowadays more than 50 per cent of waste ends up in these facilities. With that kind of management model, the more there is to burn, the better, so why try to limit the amount of waste produced? Every year, the average Danish resident produces twice as much rubbish as the average Pole, the highest level in the European Union.[86]

The best-known incinerator is Amager Bakke in Copenhagen, a site housed within an eye-catching construction with an artificial ski slope on the roof. It opened in 2017, replacing a small, inefficient installation. Five local authorities clubbed together to build it. Some wanted to create a smaller site and increase the emphasis on recycling, but that came to nothing; inertia and the incineration lobby did their work, and the new Amager Bakke, half a billion euros later, is massive. Simply too big. In order to function effectively, it has to burn such huge quantities of waste that much of it has to be imported from abroad. And in summer, the site produces too much energy, so it can't operate at full capacity. This giant incinerator has predetermined the waste management strategy for years to come, torpedoing any hope of fulfilling a climate plan aimed at reducing emissions. When you add in two serious, expensive accidents in the last few years, it turns out that the great big incinerator is sometimes little more than a great big problem.

Contrary to what I heard in Targówek, incinerators do raise environmental concerns. Despite appealing soundbites about energy recovery and waste reduction, incineration – its critics point out – turns rubbish into another form of pollution. Its by-products include toxic dusts and ash, and it leads to emissions of carbon dioxide and heavy metals as well as particularly harmful dioxins and furans. These chemical compounds accumulate in body tissue, act as carcinogens and cause serious defects in embryos. One critical report comes from Denmark, a country that, as I've said, relies on incineration. Checks carried out by the Danish Ministry for Environmental Protection have shown that, since 2014, the Norfos incineration plant has regularly and significantly exceeded permitted emission levels of dioxins, furans and other toxins.[87]

And it isn't just a matter of outdated infrastructure. In a report from November 2018, Zero Waste Europe and Toxicowatch presented evidence that the Dutch incinerator Reststoffen Energie Centrale, one of the newest installations of its kind, emits significantly more pollutants than is allowed under EU rules.[88] In May 2019, the Dutch Council of State, the highest administrative court, ruled that the site was miscalculating its emissions readings and that, after correction, those emissions exceeded legal limits.[89] Analysis of eggs (a sensitive and infallible pollution biomarker) laid by hens reared near the incinerator demonstrated high concentrations of toxic substances. Zero Waste Europe claims that sites used for energy recovery may in reality be one of the biggest sources of environmental pollution because of their large, often underestimated dioxin emissions. Many environmental organisations believe that incineration should be downgraded to the same rung as landfill disposal in the waste management hierarchy.

*

I visited a landfill site once, a good many years ago. I didn't go for the rubbish. I was just beginning work on my book *The Birds They Sang*, and I wanted to check that I still enjoyed birdwatching. We went to the dump in Gliwice, where an Iceland gull had been spotted the day before, having ended up in southern Poland by some inexplicable twist of fate. Iceland gulls, as the name suggests, are more suited to polar climes. It was January, and therefore winter, when gulls travel long distances. The Iceland gull was sitting in a many-thousand-strong flock of black-headed gulls which roam all over Europe during the winter months. Every now and then, they'd all take to the air with a scream.

Searching for a white gull amid such a throng may seem like looking for a needle in a haystack. But to the experienced observer, its rather subdued colour, like milk with a drop of coffee, stands out. The gull kept vanishing into the circling flock, but after a few minutes one of the binocular-clutching ornithologists would cry out: "There it is!" And then came the tedious attempts to pinpoint it: "At the level of the third pole to the right of lamp post . . . in the second row . . . it's just passing a herring gull . . . it's gone now." Much like its companions, the Iceland gull had come here to eat. In winter, rubbish dumps entice both gulls and birdwatchers.

The warm stink of decay belched from the long plastic pipes protruding here and there. Organic waste covered by successive layers of rubbish was decomposing despite the frost. Ignoring the gulls and birdwatchers, the bulldozers patiently compressed the mountains of bottles, cans, tyres and rags.

Is Europe abandoning landfill sites? Not entirely. There will need to be somewhere to put the leftovers from incinerators, at the very least. Europe only thinks it's abandoning landfill because it sends its waste to countries where regulations are more lenient or easier to bypass. The wealthy pass their rubbish on to the poor – they're the ones who bear the true cost of our sanitised lives.

MARCH 9, 2019

9.10 a.m., 52°08'38.9"N, 21°04'39.6"E

JULY 2, 2018

2.10 p.m., 54°49'58.1"N, 18°06'12.9"E

The balloon was grey, but it caught my eye even amid the March gloom – probably because there's very little rubbish in the gated nature reserve of the Natolin forest on the outskirts of Warsaw. It had recently landed there from someone's birthday party, and it hadn't even had time to deflate. Among the monumental oaks of these beautifully preserved riparian forests, the grey intruder stuck out like a sore thumb. The balloon on the beach, by contrast, was lost amid the hordes of other rubbish, mostly cigarette butts and various pieces of packaging. And this Baltic specimen wasn't in as good shape as its Natolin counterpart, just a flimsy shred of rubber on a long, plastic ribbon.

Balloons may look harmless and childish, but Australian researchers have shown that they are deadly to sea birds.[90] Eating balloon remnants is much more dangerous for them than swallowing pieces of hard plastic. Petrels, shearwaters and albatrosses often mistake the bits of rubber for their preferred delicacy – squid. Most balloons found inside birds' stomachs

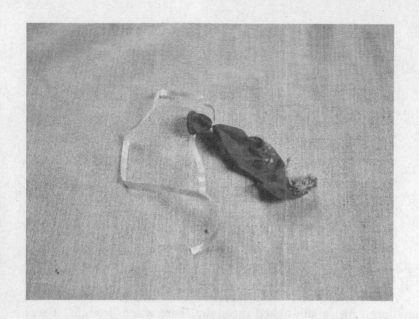

were vestiges from birthday parties, and on some you could still make out the message: "Happy Birthday".

People express their joy in numerous irresponsible ways. Spraying confetti is a popular example. These small, shiny pieces of plastic foil were fired into the air during a Gary Barlow concert in 2018, triggering protests from environmental organisations and even some angry fans. The show had taken place inside the botanical gardens of the Eden Project, where rainforest flora, among others, is housed beneath huge, transparent domes. Barlow apologised and promised it wouldn't happen again.

Personally, it's the trend for sky lanterns that I find deeply irritating. Originating in Asia, they were made popular by American films. A sky lantern consists of a paper shell surrounding a metal frame. You make a wish, light the small candle at the base of the structure and, as it fills with heated air, the lantern flies off into the unknown. It looks pretty – no wonder they've become fashionable at weddings, first communions and birthdays. But they're also lit for no reason at all. You can buy them easily at the seaside, for instance, take them to the beach, set them alight and watch as they rise into the air. And what next? Who cares? Usually, after a few seconds, the lantern drops pitifully into the water.

I've found two lanterns in recent months. The bleaker of the two was lying on a breakwater in Darłowo. The sea had given it a good battering: it looked like a sad, repeatedly used condom. I found the other one in the Owl mountains, where a spruce forest met a sun-baked meadow. There was a drought, it was even too hot for the crickets. Did the romantic halfwit who lit that lantern somewhere nearby spare a thought for what might happen if a sudden gust of wind carried it over someone's house, or over a field or a forest? Their lack of imagination depresses me.

Sky lanterns are banned in many countries – in Germany they're forbidden in most Länder. They're not allowed in Krefeld, in North-Rhine Westphalia, for example. Unfortunately, a mother and two daughters weren't aware of this when they bought a lantern online and lit it on New Year's Eve. It fell onto the ape house of the city zoo and started a fire that killed thirty animals, including fruit bats, birds, chimpanzees, marmosets, five orang-utans and Europe's oldest captive Western lowland gorilla. During the press conference, the local police chief said the perpetrators had expressed profound remorse. They faced up to five years in prison.[91] Sometimes there's a heavy price to pay for thoughtlessness.

5

THE FROG MYSTERY

The *Julodimorpha bakewelli* of the Buprestidae family is a hefty beetle that looks like a half-digested seed or a porous, glossy date. It lives in the Australian bush and feeds on the leaves of endemic shrubs of the myrtle family, with their striking red flowers. When the mating season arrives, male beetles patrol the land from the air in order to find a flightless, equally brown and porous female partner. Copulation takes a fairly familiar form. The male exposes his aedeagus, hidden under his body, climbs onto the female and transfers his sperm into her. In 1980, a photograph was taken showing a *Julodimorpha bakewelli* attempting to mate with a discarded beer bottle.

Two scientists – Darryl Gwynne and David Rentz[92] – began looking into the case. They found that *Julodimorpha bakewelli* searched for a specific type of glass bottle. The beetle was particularly fond of "stubbies" that shone in the sunshine – bottles with a volume of 370 millilitres and a characteristic ring of small symmetrical dimples around the base. Those features, and the fact that the bottles lay motionless on the ground, meant that the males mistook them for abnormally large females of their species. Giant females are more attractive than regular ones. It's like how some birds prefer to brood big, brightly coloured fake eggs that closely resemble their own. Just as the nameless prehistoric sculptor was captivated by the giant bust of the Venus of Willendorf, the beetle dreamed of a brown bottle. Scientists call an exaggerated version of an attractive feature a supernormal stimulant.

Stubbies were porn for the male *Julodimorpha bakewelli*; far more interesting, far more arousing than sex with actual females of their species. They completely ignored the real thing. The beetles wouldn't leave their supernormal stimulants unless the scientists moved them. Tirelessly they attempted to copulate, unphased by their partner's cool response. Actually, the bottles were far from cold; they caught the sunlight, and many ardent *Julodimorpha bakewelli* were fried to death, and not through fiery passion. Some males were unable even to climb onto the bottle, let alone fly away. They fell victim to *Iridomyrmex discors* ants, which attacked the beetles' sensitive, exposed genitalia. The scientists judged this to be a threat to the species and sent the results of their research to the beer makers.

The solution to the beetles' unhealthy desires turned out to be surprisingly simple. It was enough to change the colour and shape of the bottle. Green bottles didn't awaken such fervour. The beetles' world was put back on track. The scientists received the 2011 Ig Nobel Prize, a satirical version of the Nobel awarded by the *Annals of Improbable Research* for "research that makes people laugh, and then think".

A stubbie that a beetle mistakes for a beetle is an example of what has been known since the 1970s as an "ecological trap". These arise when environmental changes create a situation where a species' typical behavioural pattern prevents them from fulfilling their life's mission, that is, passing on their genes. Their biological programming, which has always worked flawlessly, can't update quickly enough and instead becomes a curse.

Such is the nature of traps – they entice, lure, promise, before suddenly becoming a deadly threat. A concrete tank that amphibians can easily jump into but are unable to leave is a trap. Common frogs found in parks may spawn in water, but they spend most of the year on land. A fishing net resembling the seaweed that gannets bring to their nests is also a trap. Chicks and adults alike become entangled and die in the indestructible plastic coils. So too is litter dropped in forests, as for various reasons it's very attractive to organisms that riffle through the undergrowth. It's easy to slide into a bottle on the ground, but it can be very hard to climb out.

Reading about whales filled with carrier bags or turtles caught in nets almost every day, you might get the impression that rubbish is only a problem in seas and oceans. But it doesn't all magically end up in the sea. It makes its home quite unremarkably in landfill sites, in woodland leaf litter or by the side of the road. Or it quickly disappears from view, buried in illegal dumps – in old gravel pits, in marshes or on private property. Bottles, tyres, food scraps and dangerous toxic waste. Fibre cement, broken tiles, mineral wool. All of this also alters eco-systems. Rubbish equals pollution: harmful substances released into the water and soil, in many cases materials that will still be decomposing when our bones have disappeared without trace.

*

This may sound like blasphemy, but there is another side to the situation – for many animals, rubbish becomes a place to find shelter, food or a mate. It can be said to provide a range of unexpected opportunities. For some organisms it represents a lucky break. Sometimes it's as if waste is rebelling against humans, its ungrateful creators, such as when it forms an alliance with our enemies – for instance with dengue-carrying Egyptian tiger mosquitoes. These insects were effectively controlled in Brazil until a few decades ago, but in the 1980s and 1990s, the problem grew worse and more complicated alongside a sudden increase in the levels of rubbish. It transpired that the mosquitoes had taken to laying their eggs in cans, tyres, bottles and plastic packaging, usually single-use items not fit for recycling, and above all in objects containing rainwater. First and foremost, the problem affected rubbish kept outside people's homes.[93]

In our part of the world, a good example is the European fire ant, which loves a polluted landscape. The ants' colonies tend to be more densely populated on illegal dumping grounds than in natural conditions. It's a rather tolerant species, settling in a wide range of environments so long as they are at least partially damp. At dumps, they live in old bottles, buried cans, scraps of cloth or carrier bags. In more unspoilt areas, they make their nests in grass and under stones and pieces of wood. But these habitats were left uncolonised in areas containing litter. Researchers believe that the fire ants' rejection of such natural sites may be linked to more plentiful supplies of food (small invertebrates) found among piles of rubbish.[94]

Birds, like insects, have long been known to make use of our rubbish. Take storks, for instance. Birds from Western Europe often settle near dumps, where they can feed and find nesting material. They've even stopped migrating to Africa – garbage provides nourishment all year round. Polish ones are more traditional and mostly look for building materials in fields, though this doesn't change the fact that half of all nests contain litter. In studies carried out in the west of the country, more than seven hundred nests were monitored over several years.[95] Unsurprisingly, the results showed that nests in urban areas contained more waste, mostly consisting of the familiar string, plastic film and pieces of cloth.

Less than 1 per cent of chicks were killed by our civilisation's waste in their nests, a number that is perhaps offset by the fact that the adult birds saved a lot of time they would otherwise have spent searching for natural materials. Meanwhile, there was an interesting correlation between the age of females and the amount of litter brought to the nest. Older, more experienced birds that had raised more nestlings were more likely to use anthropogenic materials. They didn't waste time looking for straw, now a rare sight in villages, and they didn't fly to distant woods in search of sticks. They gathered what was available, accessible and closely resembled their traditional building blocks: rubbish.

Similar behaviour has been observed in black kites, predatory birds that are also found in Poland. Older birds that have had more breeding success and occupy more fertile areas are more inclined to use rubbish in their nests.[96] Does that mean birds prefer waste? Is nature adapting to our opulent consumer habits? Or perhaps birds now consider our litter a natural part of the landscape? The answer isn't clear. Studies of black-faced spoonbills living in a highly polluted environment showed that, despite the abundance of waste, they readily used natural materials provided by ornithologists,[97] as though they could still tell the difference when they had a choice.

*

I'm a big fan of the great grey shrike, a small bird of prey from the shrike family. It's quite rare in Warsaw; I spot it most frequently in a large glade in Powsin, on the outskirts of the city, where it has wintered for several years. You can see its focused, expectant silhouette and its long tail from far away, across the gloomy expanse and the leafless field boundaries. When it's stock still, searching for its victims, the great grey shrike looks like a statuette. Its scientific name is *Lanius excubitor*, meaning "sentinel butcher" – an accurate moniker, it has to be said, if a little melodramatic. The great grey shrike isn't a torturer or perverted sadist, but it does have a peculiar habit of impaling its victims on thorny bushes that, so it's believed, serve as a kind of larder. The spikes also make it easier to tear prey into bitesize pieces. A shrike's larder is typically stocked with mice, lizards, small birds and large insects.

The great grey shrike is a real innovator, a pioneer in the upcycling of litter. It builds its nests near grassland, on the edge of forests. Admittedly, these are not architectural masterpieces, but simple, open affairs in the shape of a basket, woven from whatever can be found in the vicinity – sticks, roots and lichen. In 1985, it was discovered that great grey shrikes had learned to build nests with polypropylene string, which had first come into use in Polish fields just three years earlier. Less than two decades later, its presence in nests had become commonplace.

In field studies carried out between 1999 and 2006,[98] 98 per cent of nests were found to contain plastic string. Great grey shrikes also used plastic bags and paper – basically, anything they could find. It's thought they were trying to compensate for the scarcity of their natural building material, horsehair. A nest encircled in string is more resilient and resistant to the changing weather of early spring. This matters because great grey shrikes begin brooding early in the year. But it was also found that a significant number of birds became entangled in the string and died. Nearly 10 per cent of chicks and several adult females met this fate. How many limbs used to be amputated by horsehair, how many wings bound or chicks strangled, is something we'll never know.

Shrikes are intelligent birds, closely related to the crow family. They're extremely creative – individuals have been observed skinning toads to avoid contact with their toxins. Others are able to catch fish or crabs, even though they are more suited to open spaces. The rapid and widespread adoption of string has to be seen as another sign of their intelligence. The birds have learned to use thorns to shred it into thinner strands.

How did knowledge of this new nesting material spread? Shrikes form new pairs every year, building a nest together, so perhaps the birds learn how to use polypropylene string from one another. The account of a Czech ornithologist Zdeněk Veselovský suggests that similar behaviour can have a local range. During the war, when RAF planes were raiding Pilsen, they dropped thin strips of aluminium to confuse the German anti-aircraft radars. The great grey shrikes took a liking to this shiny confetti. Pieces of aluminium were later found in the nests of birds that lived near the town.[99]

*

Two nests have stuck in my mind. For several years, I used to spend the summer at the coast – not right on the shore, of course, seaside resorts in peak season are an ordeal for the eyes and ears. But you need only drive a few kilometres inland, so the Baltic is just a narrow blue strip above the line of the forest, before the villages become normal again and no longer resemble seasonal marketplaces. I'd go for walks around the area, though it was mostly just fields of rape. A few copses and clumps of bushes had survived, usually where the soil was too marshy to be used as farmland. What I loved most here was the dominant feature of the north Polish landscape: beautiful avenues of mature trees. My favourite was a row of linden trees leading to the sea, which were felled in 2018 during repair works on the road. How do people not feel respect for life that has bloomed and provided shade and protection for over a hundred years? Those lindens were historical monuments, the most precious landmarks in the area. They had known generations of inhabitants. But no-one stood up for them.

That avenue has now disappeared, lost without a trace. My story took place close by, in the shade of another, rather less noble avenue of trees – poplars, this time – after a summer storm. Extreme weather takes on biblical proportions in these parts and serves as useful reminders of man's lowly status. After such cataclysms, the air is cold and clear, and the road is covered in stalks, boughs and leaves. I found two empty nests torn down by the gale, a few hundred metres apart. It was July. The proprietors, some sort of field birds who would now be picking fatty rapeseeds from the ground, must have already raised their young. The nests were quite similar in size, both were fresh, carefully woven from grass and moss. One of them was lined with reddish feathers, presumably from an old pillow; in the other they were greyer. Both were encircled in some kind of polyester padding, perhaps the filling of an old duvet or jacket that was probably still lying nearby in the undergrowth.

*

For field birds, such warm, dry material is a lucky find. But for most organisms, rubbish is a tragedy, though not all waste is equal. Natural systems are so complex that nothing is simply black and white. Besides, "black and white" is categoric nonsense, a crude, human concept. This struck me when I was bending to pick up a mouldy trainer that was half buried in a russet carpet of beech leaves. Behind me I heard Krzysztof: "Leave that shoe, it'll be a good winter habitat for toads". It was a mild December, but cold-blooded amphibians can barely handle even slight frosts. Toads need a safe, warm place in which to hibernate until spring. I left the trainer where it was.

I found Krzysztof Kolenda, or rather his profile "Chronimy Lokalną Przyrodę" [We protect local nature], on Instagram. Contributors to the page collect waste and examine what it contains. In a small lecture room at Wrocław University, I learned that Krzysztof is a herpetologist by training and wrote a PhD about the migration and mortality of the common toad on our nation's roads. About those hundreds of tyre-flattened corpses strewn across the motorways every spring. The convergence of amphibians and waste is one that has fascinated him since high school. It all began during the moor frog spring mating season at a pond in Ostrów Wielkopolski. The blue-tinted males were croaking wildly, their unearthly, almost lunar skin a shimmering azure. But the tonnes of rubbish discarded by local residents made it very difficult to contemplate nature. Krzysztof organised an annual litter pick and the local council chipped in to help. That's when he noticed the insects rattling about in the rubbish. Practically every bottle was something's home or tomb.

Towards the end of his university studies, he decided to take a more rigorous approach to the issue. He received a grant for an educational project, and with the help of pupils from his former high school he began picking up litter in a more official capacity. They collected over two hundred and fifty bottles and cans in local forests. Inside a third of them, they found living animals, and in half they found remains. Close to one and a half thousand organisms had died in the bottles. As expected, the majority were invertebrates, beetles – earth-boring dung beetles, to be precise – that whole army of organisms that cleans up the forest floor. More surprising were the remains of as many as six species of mammal. There were wood mice and field mice, a red-backed vole, a fairly large field vole, and shrews – a common shrew and a pygmy shrew.[100] [101]

It was only a small, semi-amateur initiative, and yet it threw up so many interesting finds. The Polish Press Agency[102] published an article and the "Teleexpress" news programme recorded a short segment on it. Following this wave of success, a journalist from a local TV channel turned up. She was excited to see the collection, but when they showed her the cases of pinned specimens, she couldn't hide her disappointment. "But these are just beetles," she pointed out perceptively. The finds weren't impressive enough for her. This story makes me laugh, although to be fair, when I hear the word "animal", I too picture something at least the size of a hedgehog. We're so self-absorbed. From our perspective, the lives of small beetles seem meaningless.

*

Some time ago, a photo of an emaciated wolf with a plastic jar on its head was widely shared in the media. Then came the Slovakian bear in a metal bucket and the deer in Wrocław with its head in a plastic bottle.[103] Large, charismatic fauna, completely helpless and condemned to death by our thought-lessness – these situations enrage us and stick in our memories. But who cares about a few hundred dung beetles trapped in a discarded engine-oil canister on the hard shoulder? The fate of invertebrates or small rodents imprisoned in our waste is a niche concern. Clearly, when someone decides to study biology, they dream of researching the social behaviour of wolf packs, the diet of bears or the cognitive abilities of primates. As chil-dren we all watch the same nature documentaries, and few are born with a passion for phytoplankton or bryology (the study of mosses). The appeal of a springtail that decomposes dead organic matter is rather more subtle and obscure, appreciated only by a select few.

The first article about the influence of discarded waste on the animal kingdom was written several decades ago in the UK. The world was still relatively clean, almost devoid of plastic. "In 1961, it was noticed that a milk bottle lying in a hedge contained the remains of eight small mammals," reads the introduction. How come they couldn't get out? It turned out that the number of remains found in a bottle depended on its size, its position and the size of neck. Many animals aren't able to squeeze their way back out when all they have beneath their feet is slippery, often wet, glass.[104]

Those British scientists didn't inspire many people to follow in their footsteps, though the amount of rubbish was growing. Over half a century, barely thirty papers were written about organisms trapped in waste, and some of them could more accurately be labelled as notes rather than articles. Most concerned mammals, which are significantly more interesting for researchers. Obviously, a springtail is no chimpanzee, but there are also purely technical reasons for this narrow focus: mucking about in a black soup of limbs, shells, larvae, pupae and other remains that can't be named or properly counted is a rather thankless task. Therefore, only in around six papers did anyone attempt to identify invertebrates.

Back in Poland, studies carried out by Jarosław Skłodowski and Wojciech Podściański in the Tatra mountains involved intentionally laying down various beverage containers,[105] so they could measure how quickly different creatures settled inside them. After four weeks, the bottle from a drink given the symbolic label "cola" had caught most organisms; after eleven weeks, beer bottles had outstripped the competition. Flies came first, lured by the sugary scent, followed by, among others, predatory insects and detritivores, which feed on dead remains. Today this methodology is considered unreliable (litter lies in forests for years, not weeks), not to mention controversial. Dumping rubbish in a forest, even for research purposes, is not exactly an admirable pursuit.

*

Krzysztof Kolenda took the opposite approach: rather than depositing waste, he focused on picking it up. Though he lacked any financial support, and was in the throes of a PhD, Krzysztof began his own guerrilla research project, cleaning up the forest and studying the contents of cans and bottles. He wanted to show that it was possible to do something good for one's environment and for science at the same time. And it really makes a difference – around fifteen years ago, research showed that regular litter picks reduce the mortality of small animals. It was calculated that if litter collection ceased in Poland's Tatra National Park, thirty-six million invertebrates and ten thousand small vertebrates would die in bottles or cans every year. That might well have an impact on the equilibrium of the entire ecosystem.[106]

Krzysztof's grassroots project attracted a squad of specialists from various disciplines. They selected ten forests (perhaps "small woods" would be more appropriate) in Wrocław and set out to collect about a hundred open bottles and cans from each one. The stunt failed in just two places. They looked for rubbish within ten metres of the path, roughly the distance a casually flung bottle will travel. They carefully recorded the material, colour, original contents, volume, the diameter of the opening and the presence or absence of a neck. Above all, the team wanted to find out what kind of rubbish makes the most effective trap, and with such a large sample, some interesting discoveries were to be expected.

*

Back in 2018, I was invited to view the collection. The contents of almost six hundred bottles and cans fits inside one cardboard box, a tomb for thousands of creatures. Shells of predatory ground beetles, true tigers of the forest floor, floated in a thick ethanol soup, matte and dark like toasted sunflower seeds. Over time, 70-per cent alcohol turns everything black and erases its shine. From time to time, I caught the amber flash of a sexton beetle's wing case, waiting in plastic containers to be analysed. Larger finds are kept in pots that look like they were designed for urine samples. Smaller ones end up in little test tubes known as Eppendorfs.

There was one reptile: a young slow worm, dry as dust, lay in pot 938. Among the larger remains there were also small mammals – mostly shrews, whose skulls can be identified by the very thin bone casing, and whose jaws are studded with characteristic reddish teeth. Shrews use bottles as a kind of burrow, squeezing through the neck with genuine pleasure – such narrow tunnels clean their fur. And they're certainly not indifferent to the smells coming from inside them, either. The deadliest bottle that Krzysztof found had a volume of 1.7 litres, a diameter at the opening of 19 millimetres, and contained the last drops of a sugary drink. It housed the bones of sixteen rodents.[107] But the world record is fifty-four small mammals in a single two-litre water bottle.[108]

It seems some animals use rubbish as a "safe" hiding place. Several glass bottles were found to contain stocks of seeds, probably brought there by rodents. That kind of bottle makes for a decent larder – especially if it's lying in a dry, shady place with the neck pointing towards the ground. Plastic bottles, on the other hand, are particular favourites of spiders. Almost one in three had traces of cobwebs and half contained shed skins and living or dead specimens. But the most surprising thing was that several spiders, often of different species, were found inside a single plastic bottle, when normally they can't stand to be around one another. Perhaps there was enough space in their shared hideout to avoid any pushing and shoving?

There were many mysteries to be solved. What were molluscs doing in a bottle found in a roadside ditch? Hypothesis: the ditch periodically filled with water, and the larvae may have been carried there on the feet of birds. They tried to find a safe harbour in the bottle (at this stage in their development they can swim), but when the ditch dried out, the molluscs died.

And what were ants doing in the molluscs' shells? At first the researchers thought it was a coincidence. But one expert in the team, a myrmecologist, told them that ants of the *Temnothorax* and *Leptothorax* families are known to colonise empty snail shells. Apparently, they prefer the shells of Cepaea land snails, those familiar stripy creatures found in practically every garden in Poland. And, on the subject of shells, among the true jewels of the collection are the intricate, conical homes of the door snail, resembling miniature Angkor Wats.

But the most mysterious find was the frog. A dead, adult brown frog trapped in a bottle. The age of a frog can be determined very precisely: the bones are decalcified and dissected to reveal rings that mark what is called the period of arrested growth, one for each year of hibernation. Bone analysis – or skeletochronology, as it's known – showed that the frog was six years old.

How did a six-year-old frog end up in a beer bottle? It was definitely too big to climb inside it, but it's hard to imagine it lived its entire life in there. One clue was the fact that the skeleton was incomplete. So, in all likelihood, some other creature had dragged the frog's remains inside and eaten them at leisure in its glass retreat.

A Nokia FM210 receiver. 85 millimetres long, 60 millimetres wide – roughly the size of a credit card. Thickness: 22 millimetres. Weight: just under 150 grams (not including the clip) – the same as a cup of flour. I'm including these details because I think we've more or less forgotten what a pager looks like. Nowadays, an SMS contains up to one hundred and sixty characters. A message on a pager could hold sixty-eight. Concise thoughts, pure, uncontaminated meaning. And capacity for up to sixty messages. No lifelong archive.

These are the figures for just one of several models that I found among dozens of devices dumped on the fringes of a Mazovian birch wood. Electrical components, motherboards and chargers lie scattered among them. They don't look as though they've been here for twenty years, since the final days of the pager's era. Someone threw them out recently. Perhaps they discovered the devices in their basement or garage? Maybe it was the owner of a defunct telecoms shop? In theory, you can find out who owns these pagers – each has an individual serial number. In Poland, the Polpager system had barely forty thousand users. The devices came on sale here at the same time as the change of political system, but their moment in the sun was short lived. By the late 1990s, mobile telephones were becoming more widespread.

The first pager was patented in the middle of the twentieth century. It received radio waves that carried an alert about an attempted telephone connection. The caller's message could be

listened to via a call centre. Over time, pagers were improved so they could receive text messages and even send them. They reached the height of their popularity in the 1980s. In 2017, it was found that one in ten of the world's pagers belonged to the UK's National Health Service.[109] Maintaining the devices is expensive, but the advantages of radio communication become clear in crisis situations when mobile networks are overloaded.

A naturalist friend of mine told me about an illegal dumping ground in an old gravel pit near Osieck. Guided by some kind of atavistic instinct to bury waste, people come here to throw their unwanted items into the quarry. The landscape is typical for this part of Poland: sand, feeble young birches, rows of planted pines. And rubbish. It's mostly bulky waste that ends up in illegal dumps, cumbersome clobber you can't just throw in a bin. Like an old television with a smashed picture tube. Strontium, barium and lead seep straight into the soil on the edges of the Mazovian landscape park. Buried beneath the leaves, with vegetation weaving its way through it, is a layer of old carpets and clothes and sacks of polystyrene from construction projects.

It's easy to find a place like this in practically any suburban forest. It'll be discreetly concealed, always out of the way, but the local residents will know it well. Whichever way you look at it, throwing waste into the bushes is a shameful activity, though a very common one judging by the amount of rubbish you find. It reveals a kind of misguided desire for tidiness –

after all, illegal dumps generally operate in the same place for years. Like real landfill sites, only free. Spoiled landscapes have become so commonplace that a fridge rooted into the forest floor seems almost natural.

6

THE LIGHTBULB CONSPIRACY

Apparently, I'm not the only one to keep a small museum of progress inside a drawer. Coils of useless cables, stray chargers, oddly shaped batteries. Does anyone remember the 3R12? I threw one out quite recently. It only fitted in a plastic torch my godfather once brought me back from England. The torch was old and square – nowadays it would be a fashion statement, but when I took it on a camping trip at the age of twelve, I felt stupid. All my friends had new rubber Patrol torches, cheap substitutes for the police Maglites from American films. I felt as though I were wearing a pair of unfashionable, hand-me-down shoes with holes in them.

The torch my godfather gave me came from prehistoric times when devices worked for ever, you only needed to change the bulb and battery. Old mobile phones, which take pride of place in my museum, also seem like relics of a long-bygone age. Thick and heavy, as unwieldy as old computers, typewriters or telephones with dials. Protruding, childish buttons and a ludicrously small screen. My first phone, from high school, was an Alcatel with a pull-out antenna that snapped after a month and a keyboard hidden beneath a flap. It could hold fifteen SMSs. Deleting messages like "Key under doormat" or "Soup in pot" was wholesome. Nowadays we're constantly carrying that grey mass of life's mundanity around in our pockets, an archive of quotidian concerns.

Next came a slightly smaller Alcatel with an external antenna that snapped after two months. It boasted polyphonic ring tones and a slightly bigger memory, though it still had a black-and-white screen with a two-line display. By that time some people already had phones with cameras, which took dark, blurred pictures the size of a postage stamp and with a one-second delay. Then came an anonymous parade of Samsungs, Motorolas and Nokias, before the average phone started shamelessly pretending to be at once a computer and a camera with a failsafe algorithm and a brilliant lens. It didn't pretend to be a watch because it was one – for the last decade only politicians, presidents or company directors have worn these archaic objects on their wrists. And perhaps people who've been left a priceless heirloom by their granddad.

I still wonder why I keep that jumble of gadgets and cables – including a Dictaphone I borrowed from an ex-girlfriend twenty years ago. Why did I hang on to that 3R12 for so long? Was it a vague feeling that it might come in handy one day? Or a sense of wartime prudence inherited from my grandmother? Perhaps it was for the same reason I bring a stack of unread books with me every time I move house. A strange, hard to describe sentiment that basically means I crave the new but feel bad about the old. So, while I still have space, I hoard. But what happens when the space runs out?

Have you ever noticed how more and more "self-storage" warehouses are springing up on the outskirts of towns? Don't have time to go through the things in grandma's old apartment? Throw it all into a clean, dry room and never think of it again. Anyone who's ever watched cable TV will have seen an American programme where they auction off the contents of abandoned storage units and garages. They might conceal trash or treasure, smelly old furniture or collectors' baseball figurines from the 1950s, which someone will buy for thousands of dollars. Or a gun carriage from the Civil War.

*

According to a report by the World Economic Forum, in 2018 humanity produced 50 million tonnes of electrical waste. It is the fastest growing type of waste, increasing by between 3 and 5 per cent every year. The total value is estimated at 62.5 billion dollars annually.[110] Unfortunately, barely 20 per cent is recycled. The rest ends up in landfill sites and incinerators, or is crudely dismantled in the open air. We have no idea what happens to three quarters of all e-waste.[111]

The global market thrives on growth, on constant increases in turnover. That's why electronic equipment is sold with ever-diminishing lifespans. Premature death is artificially pro-grammed: things are designed to break at the end of the guarantee period. We've learned to just stoically shrug our shoulders, but the situation is changing. From 2021, in the EU and UK, manufacturers will be obliged to produce longer-lasting appliances and to supply replacement parts for at least ten years. These regulations, adopted at the end of 2019, concern washing machines, dishwashers, fridges, televisions and lighting. Critics say it's an imperfect solution, because consumers won't be able to carry out the repairs themselves, while the workshops that were commonplace just a few decades ago have closed their doors, letters dropping one by one from the signs above the dirty windows: ELECTRONICS REPAIRS & SERVICING.

But for now, the old system lives on. Zygmunt Bauman has gone as far as saying that wastefulness is written into the DNA of consumer society. Our economy only hangs together when it's accelerating. As it grows, the pantheon of products nobody needs expands with it. The manufacturers' task is therefore to create new needs, keeping consumers in a state of constant unfulfilment, with the feeling that somewhere out there, something better, newer and more expensive is waiting for them. Old things start to feel like a drag, we have to throw them out to make room for the new. In this sense, buying is replacing. The faster you can make a consumer feel dissatisfied with what they have, the better. A sated, gratified customer is no good to anyone.

A refrigerator is a relatively simple appliance used for cooling food. That's all. It takes a lot of brainstorming to come up with new functions and to persuade buyers of their utility. A few years ago, a fridge was marketed that let you see what was inside just by tapping on the door. Someone correctly pointed out that every time the door is opened, energy is lost and the appliance has to work harder to maintain the desired temperature. Yes, this is obviously true, but what about the cost of producing a fridge "with a window"? Can the screen be removed? Can it be properly recycled? Will any treatment facility be able to handle it?

Bauman writes that the customer–product relationship has encroached into the world of interpersonal connections. Here, too, we want unlimited freedom of choice, freedom of action, we're not interested in concessions or compromises. Perhaps that's precisely why our role as consumers is so attractive: we feel comfortable in the technological world because it's easier to reach an agreement with an electrical device than with another person. In a relationship with a product, we're always the subject, never a docile object. You don't have to negotiate with a purchase, you don't have to make sacrifices or live with the feeling that things aren't going as well as you'd hoped. A new washing machine, fridge or curling tong will read our minds, it will be a silent, trusted friend and helpmate. In this relationship we don't have to be faithful. When we tire of things or when they stop being useful to us, they go straight to the dump without a second thought. Their time runs out ever more quickly.

*

How is it possible that selling products with a pre-programmed lifespan is even allowed? It's by no means a new idea. The first product to be designed with planned obsolescence in mind was the lightbulb. In the late nineteenth century, a bulb would usually shine for two thousand hours. At the start of the last century, they lasted even longer. And yet, nowadays, classic lightbulbs reach one thousand hours at best. This came about because of the lightbulb conspiracy – bear with me, I promise that isn't some outlandish theory I read about on the internet.

Imagine a dozen or so respectable moustachioed gentlemen, representing Phillips, Osram and General Electric, who devise a plan to make themselves a little more money. In late 1924, they establish a cartel supervised by the Swiss firm Phoebus (the Phoebus Cartel). Its objective is to inspect the production of bulbs, though it's a curious brand of inspection. The cartel ensures that the quality of the product is diminished and that sales increase. Any company whose lightbulb shines for too long incurs a fine. As a result, the lifetime of a lightbulb is reduced by more than half.

Nowadays it's hard to find products that *don't* break after a few years. Sometimes, in some forgotten countryside larder, you still come across working Minsk, Mors or Donbass refrigerators. They're simple, bursting with CFCs and ravenously power-hungry, but they're also virtually indestructible. Electrical appliances now have labels showing their energy class and adverts often tell us that new equipment consumes less electricity. Manufacturers assert that our purchases are good for the environment. But nobody mentions the fact (and why would they?) that in many cases significantly more energy is used in the production process than in the appliance's entire life cycle, at least when it comes to equipment produced in the last fifteen years or so. Manufacturing a new washing machine comes at a real cost – not just to our pockets – which all of us bear.

*

I resisted acquiring my first smartphone for a very long time. I would proudly show off my small, ungainly, ancient models with no internet connection. I was afraid I'd fall into the trap of constant availability, responding to emails day or night and double-checking every doubt on Google. Three years ago, I gave in. My girlfriend persuaded me, she said it would make my life simpler. I also thought – though this may sound silly – that I'd be able to input birdwatching data straight into online ornithology records. For instance, I'd tap in "red-backed shrike, one specimen" and the GPS would correctly locate me next to the Palace of Culture.

As you can see, there may be a variety of reasons to buy a smartphone. But, nonetheless, my worst fears were realised: my device – that modern symbol of the lower-middle classes – took control of my daily life. If I found myself with nothing to do for a moment, I'd unconsciously start swiping through flickering pictures, news, or the breakfasts, lunches and dinners of my Facebook friends. I'd reply to emails. I was no different from the people I'd looked down my nose at just a week earlier. I stopped reading books, especially on public transport. I was constantly checking pointless things, but it all seemed tremendously useful.

You're probably bored by these clichéd confessions of a

technologically retarded ornithology enthusiast. Unfortunately, I have a serious tendency to procrastinate, so any activity that distracts me from work quickly ends up blacklisted. These devices, after all, are designed to absorb us for hours on end. Just think of the way children hysterically demand their parents' phones. Life did become simpler, but I had to pay for it. To be on the safe side, I bought an old-fashioned phone with buttons and no internet connection. I use it day to day, though writing text messages is an ordeal. On the other hand, it still works, even though I've dropped it dozens of times, and the battery lasts for several days. My smartphone, meanwhile, mostly comes in handy as a sat-nav in the car. I also take it on longer trips instead of a laptop.

Not everyone has such an obsessive attitude towards their time as I do. The smartphone is an extension of the modern human's hand. It supplants tools like calculators, watches, alarm clocks, radios, cameras, video recorders, GPS locators, torches, spirit levels, compasses and even magazines in our daily lives. We start our day by checking our inbox, social media, the weather or pollution forecasts. The smartphone is the confidant for our secrets, it manages our private life, work and everything in between. It's a long time since it was considered shocking to see people gawping at their phones on public transport first thing in the morning. They scroll through news, they play games, they shop. It'd be hard to convince them to give up such a universal and useful device for any reason at all. It seems we can no longer live without them.

Increasingly, for citizens of the West, a synonym for a good rest is a "digital detox" in a place beyond the reach of mobile networks. There are fewer and fewer such places. Over eight billion SIM cards are in use in the world, in the hands of five billion people. Every year, more are sold and records are broken. In the years to come, hundreds of thousands of people in developing countries like India, China, Pakistan and Bangladesh will buy their first mobile phone. Estimates predict that in 2025, nearly six billion people will own a mobile phone.[112] That means 70 per cent of the global population will be using them. In 2018, fifty-five new models were produced and one and a half billion phones were sold.[113] It's estimated that a device bought in 2018 will be replaced for a new model in just under three years. Older data suggests that we replace our phones every eighteen months.

The cheapest smartphones now cost around a hundred pounds. They're obviously second-rate and the moment their owners leave the shop, they're already thinking about replacing them for something newer. But the fact that phones don't cost much doesn't mean they aren't valuable. Processing smartphones is the most lucrative sector of e-waste recycling. A modern phone contains thirty elements from the periodic table. Metals like silver, platinum and gold are used as conductors in electrical contacts. The iPhone 6 contained barely 0.014 grams of gold, an amount that is worth a little more than fifty cents, but which constituted half the value of all the elements found in the phone. That's not much, but it's worth remembering that a tonne of concentrated raw material from recycled phones contains one hundred times more pure gold than a tonne of ore extracted from the earth. Close to 70 per cent of the iPhone 6's mass was made up of more prosaic elements such as aluminium, iron and carbon.[114]

Rare earth metals, also used in the production of electric cars and batteries, provoke the most heated reactions. It's not too great an exaggeration to say that our future depends on how we deal with their extraction. Many of the elements that are essential to our phones had no practical use just a few decades ago. Ytterbium, for instance, which is now used for building lasers, was once considered a mere chemical curiosity. For the most part, these metals aren't all that rare, but acquiring them is very complicated.

First of all, the metal deposits are often in places that are very hard to reach. Second, they are difficult to separate. They have similar atomic weights, melting points and electrochemical structures. Third, extraction often takes place in violation of human rights and at enormous environmental cost.

So why isn't repairing devices encouraged? Why are we bombarded by incentives to buy new, ever more advanced things, while useable equipment ends up in landfill sites? Why are second-hand products, which could be sold at competitive prices, not promoted in the same way? Businesses are fixated on growth – from their perspective it's less profitable to fix old items or replace worn out parts. What matters most is accelerated production capacity, expanded logistics and the constant flow of money to successive links in the chain. But why does our world take the side of producers, rather than consumers? And why aren't we bothered by the enormous cost to the environment of our obsession with ownership?

*

If barely one fifth of electrical waste goes to appropriate, specialised treatment facilities, what happens to all the rest? It ends up in landfill, or it's crudely recycled with no respect for any standards. Every year, mountains of e-waste are sent to countries where their subsequent fate goes undocumented. To countries where regulations for recycling such waste do not exist or are not enforced. One of the most famous e-waste dumps is Agbogbloshie, a deprived neighbourhood in the suburbs of Accra, the capital of Ghana.

Half a century ago it was marshland. Now, inhabited mostly by economic migrants, and known as "Sodom and Gomorrah" thanks to the rampant crime, it presents a post-apocalyptic landscape: heaps of monitors, charred computer casings and broken motherboards stretch as far as the eye can see. Photocopiers, gap-toothed keyboards, smashed picture tubes. The dump is shrouded in black smoke from the fires in which the equipment is "processed". Adults and children dig through the piles of rubbish, cows graze stoically here and there. It's likely that waste from every corner of the globe ends up in Agbogbloshie, but it's very hard to establish how much comes from each country.

E-waste comes here as "used" or "second-hand" electronics because shipping hazardous waste across borders is prohibited by the Basel Convention. It all started in the late 1990s, when old computers and televisions were first sent to African countries from the West. The intention was to plug the technological gap between developed and developing countries, but according to estimates up to three quarters of the devices no longer worked.[115] The only resources recovered here are metals such as copper and aluminium, though huge quantities of lead, arsenic, cadmium, dioxins and furans are released into the air, soil and water. Research by IPEN (International Pollutants Eradication Network) and the Basel Action Network (BAN) has found that the level of dioxins in one egg laid by a hen roaming freely in Agbogbloshie exceeds permissible limits by 220 per cent.[116]

BAN tracks our waste. A 2018 write-up of a two-year investigation casts some light on what happens to electrical waste from several EU countries. Three hundred and fourteen old computers, printers and monitors were fitted with GPS trackers and handed in to collection points. The results were crystal clear: nineteen devices were exported outside the Union's borders, of which eleven went to developing countries like Ghana, Nigeria or Thailand. On average, equipment that left the EU travelled over four thousand kilometres. It's estimated that around 1.3 million tonnes of undocumented e-waste leaves EU countries each year.[117]

Similar research into e-waste exports was carried out earlier in the United States. The conclusion? As much as 34 per cent of discarded devices leave the country. The United States hasn't ratified the Basel Convention, so the process flourishes in compliance with the law. According to BAN, 6 per cent of e-waste leaves EU countries, among which the United Kingdom was the biggest exporter while it remained a member state. Trackers fitted inside LCD monitors (containing cadmium) went to Nigeria, Tanzania and Pakistan. Polish rubbish was also checked, with twenty trackers planted. One of them sent its final signal from Volochysk in Khmelnytskyi Oblast in Ukraine. BAN believes the transport was "likely illegal". The cathode-ray-tube (CRT) monitor dropped off at SWWH Wojciech Hanc Secondary Raw Materials in Michałowice near Pruszków was broken and could not be repaired.[118]

*

The Polish system for the collection and treatment of e-waste is not, to put it lightly, a model of best practice. Like all sectors of the waste industry, it is drastically underfunded. Poland is a paradise for producers of electronics. Charges for placing new devices onto the market are scandalously low, which is why the entire system is floundering. First and foremost, the collection system does not work. Each of us ought to take our e-waste to designated recycling centres (known as PSZOKs in Poland), but this requires time and a modicum of good will. Warsaw's PSZOKs are located far on the outskirts of the city. Who can be bothered to plough through traffic jams to deposit a broken radio? To be fair, there is a mobile recycling point in my district, but my neighbours still throw their defunct radios, fans and tape recorders into the mixed waste.

As usual, education is lacking – people simply don't know

what to do with their faulty equipment. They have no notion of the environmental costs of casually discarding a mobile phone. Masses of e-waste circulates in the informal economy. Properly recycling electrical devices comes with a price tag – it's much easier and cheaper to take an old washing machine to a scrap-metal dealer who will recover only the metal parts, discarding the rest. The procedure is illegal but widespread. Old screens, printers or blenders left next to bins are also collected by "private entrepreneurs", that is, people who melt the cables and plastic casings in nearby bushes.

Robert Wawrzonek, who runs the well-regarded company Remondis Electrorecykling, weighs his words carefully. He doesn't want to talk about the sins of others – people in the waste industry are careful: the competition is stiff and firms don't always play fair. May I see inside the facility? No, I may not. So I only pass by the warehouses, crammed full of sacks of ground plastic. Somewhere deep inside, in the darkness, I can make out old plasma televisions, while a thoughtful man sweeps up pieces of a speaker, a smashed screen and some cables. Robert Wawrzonek doesn't want to speculate how far official data on e-waste diverges from the truth. Everyone knows it does. We only reach current levels of e-waste recycling thanks to what's known as "making receipts".

In a nutshell, this boils down to the following mechanism: a recovery organisation pays for the collection of two hundred tonnes of waste, the facility collects and treats one hundred tonnes but in the documents everything tallies. It overstates the mass on entry and exit from the site. The difference is quite easy to conceal, especially in large companies that cover the whole waste chain: collecting, dismantling, treating and recycling waste.

No-one will put it this bluntly, but there's a sort of non-aggression pact in force today concerning many device manufacturers, recovery organisations, treatment facilities and recyclers, as well as state supervisory bodies. No-one really probes into what actually happens to waste, as long as the paperwork is in order. The authorities don't tend to be very inquisitive. The most important thing is that we are seen to be complying with EU standards.

The current system has been evolving since 2005, when the first law regarding waste electronic equipment was adopted. That's when the symbol of a crossed-out bin began appearing on new devices, which means that they shouldn't end up in municipal waste, that is, in normal, day-to-day rubbish. Under the regulations, discarded equipment should be properly recycled, with raw materials recovered and harmful substances neutralised if they can't be re-used. Each device contains easily accessible metals, such as nickel, tin, iron and copper. Recovered resources are a massive saving for the environment. The metal is already processed and concentrated, it doesn't generate as much waste as metal freshly acquired from ore, and the costs of extraction are also saved.

*

E-waste is problematic because it covers a wide range of products and is therefore difficult to treat. A watch and a large fridge are both e-waste, as are headphones and washing machines.

There are highly specialised facilities that only deal with a single type of device. Many appliances must undergo technologically advanced processes if they're to be recycled. Refrigerators, one of Remondis' specialities, must be treated in a hermetically sealed installation to prevent the release of harmful substances like Freon in old models.

There need to be systems in place to purify the air and recover the refrigerant, which must be neutralised. In the past, there were no such requirements, it was basically legal to hack a fridge to pieces with an axe. Most parts at the back of a refrigerator are scrap metal – compressor, engine, grill. Plastic and glass shelves can be recycled as well, as can any oil recovered from the engine. The inner panels are usually polystyrene, the casing metal or plastic. Insulation is made of polyurethane foam, which once separated from the refrigerant, can be used as an alternative fuel or sorbent.

We shouldn't overlook the fact that there have been certain changes for the better in recent years, Wawrzonek insists. A 2015 update to Polish legislation brought in some sensible regulations. E-waste was divided into six groups rather than ten. The generic term "bulky waste", which used to include both ovens and refrigerators, was replaced and technology-based categories were introduced. A refrigerator, significantly harder to recycle than an oven, was classified as a cooling device. This made the statistics for the collection and treatment of both types of appliance more meaningful. It's no longer possible to meet refrigerator targets with ovens, which consist almost exclusively of scrap and don't require sophisticated technology to dismantle them into component parts.

For some time, there's been talk of a state agency that would take over the work of private recovery organisations, forcing manufacturers to become more involved in recycling. As it stands, companies pay peanuts and wash their hands of the problem. The prices for recovered materials are often so low that honest treatment facilities, functioning on the limits of profitability, have to lower the quality of their services in order to keep up with the competition. Of course, there are many fraudsters out there. Sometimes all it takes to expose rogue companies is to look at their documentation – how can a facility process a few thousand tonnes of waste each year if its premises amount to no more than eighty square metres? Where is the forecourt for collected waste? Where are the treatment halls, the workshops where devices are dismantled? Where are the storerooms for pieces awaiting recycling? Or for parts that can be treated or even re-used? And, aside from everything else, how is there still space for an office?

*

Rare earth metals. Many of them, used in great quantities for the manufacturing of modern electronics, are mined in the Democratic Republic of the Congo, a country entrenched in an unending armed conflict in which more than five million people have died since the mid-1990s. This was the subject of a 2010 documentary, *Blood in the Mobile,* by the Danish director Frank Piasecki Poulsen. The film-makers descended into an underground mine where, in horrific conditions, thousands of modern-day slaves were chipping away at the rock with ordinary hammers. That's how coltan is extracted, for instance, the metallic ore from which we derive tantalum, a crucial element in today's electronics.

The mines were controlled by dozens of militant groups in turn – government soldiers, guerrilla units and armed bandits, whose modi operandi were all largely the same. Those working in the mines were at the mercy of whoever was commanding the area with the aid of a Kalashnikov. The miners, often children, were robbed numerous times over. They were taxed on their work and unable to leave the territory of the mine without paying for protection. In effect, they were slaves, stripped of their human rights. For the multinationals, intermediaries and leaders, the endless war is very profitable. Mining in conflict-ridden areas means lower prices for raw materials, and zero standards, official checks or supervision.

The film-makers tried to ask Nokia, the market leader in mobile phones at that time, where their rare earth metals came from. Nokia had always vaunted their enviable working conditions and high ethical standards. But on this occasion, after months of being brushed off, all Poulsen received were assurances that the company was making every effort, that it was working on systemic solutions, that it was running wide-ranging campaigns. Fine words, to be sure . . .

In 2010, Barack Obama signed the Dodd-Frank Act, which was supposed to protect the market from another financial crisis, but also (in a subparagraph) requires electronics companies to carry out audits on their supply chains and disclose where the raw materials used in their production processes are extracted. Minerals that are to be transported need to be registered and labelled with numerical codes. Naturally, the black economy has found its way in; illegally acquired raw materials from unknown sources are mixed in with those from certified suppliers. In practice, for a Western company to be completely certain that their products are made from ethical materials, it would have to hire people to monitor the entire extraction, processing and transport chain. That's impossible, of course, so corporations focus on making sure it all adds up on paper. In that, at least, they're quite successful.

*

One attempt at an ethical revolution in the world of mobile phones was the Fairphone, the first device made exclusively of materials from verified, ethical sources. They were designed to last much longer than other phones on the market, and the company published its full technical specifications and produced spare parts to facilitate repairs.

The first Fairphone was released in 2013. The tin and tantalum in it came from certified mines in the Democratic Republic of the Congo, but the phones were assembled in China. Clearly in the age of the global marketplace, ethics have their limits.

Two years after the premiere of the first model came the Fairphone 2. It was bigger, more powerful, more expensive, and the raw materials included certified tungsten from Rwanda and gold from Peru. The Fairphone is the first mobile phone to receive ten out of ten from ifixit.com, a website that assesses the reparability of electronic devices. But in 2017, the company stopped updating the software for its first model, and stopped making spare parts for it, too. Once again, noble ambitions were frustrated by the market. The company wasn't able to place orders large enough to meet manufacturers' requirements. And though parts for the first Fairphone can still be found on the second-hand market, not all users were happy with how their devices worked, in any case.

Experts predict that the next wave of e-waste will strike soon, when the world makes the shift to 5G technology. I once came across research showing that we don't actually replace phones out of vanity or the mindless pursuit of novelty. We replace our phones above all because they break or because they're rendered obsolete by the relentless pace of technology. They get old. This wasn't always the case; no researchers were studying me twenty years ago, but I clearly remember how I'd look forward to the end of my contract with a network operator so that I could buy a new model for one pound. But those models were primarily for making calls, they didn't differ much from one another and they were pretty much indestructible. Modern devices are more like computers – fragile, intricate and quickly superseded.

Big companies assert that they're doing everything they can to ensure their products work brilliantly, without hitches and for a long time. You'd be forgiven for thinking that in their endless generosity, they've given up on making a profit.

Meanwhile, in autumn 2019, the Italian competition authority fined Apple ten million euro for deliberately slowing down older iPhones. Users weren't informed about the possible consequences of the software updates – which often introduced bugs – nor given advice on how to preserve battery life. The Italian authorities ruled that Apple's behaviour forced users to buy new phones. Samsung, in turn, had to pay five million euros, though the company rejects all accusations that it intentionally slows down its devices. Two new Samsung models (released in 2019) were studied by ifixit.com and given two and three points respectively on the ten-point scale of reparability. By way of comparison, the company's phones from 2011 scored between six and eight points.

Apple has faced further fines over the slowdown controversy in the US and France, while a case in Israel is still ongoing. In 2018, the company introduced Daisy, its very own robot, capable of dismantling two hundred iPhones per hour. That's 1.2 million smartphones annually.[119] That same year, the American company sold 217 million iPhones.

*

The capital of Belgium is not just the seat of the institutions of the EU, but also the base for many non-governmental organisations that seek to influence Union policy. One of those is the European Environmental Bureau (EEB), which brings together nearly one hundred and fifty organisations working in environmental protection. Among its concerns is a push for controls on the use of cadmium, a toxic raw material that accumulates in living organisms and damages the nervous system. The EEB also raises awareness of the need to produce energy-efficient electronics that can be repaired and easily recycled.

Data shows that 77 per cent[120] of people would rather fix their devices than replace them. But repair costs are often too high. The EEB therefore started a campaign for the right to repair, advocating for new laws to limit the generation of e-waste. What does that mean in practice? Above all, it's about changing the way devices are designed so that they are easier and cheaper to repair. The EEB fights to ensure spare parts are widely accessible and are supplied in easily exchangeable modules.

Of course, the spectre of reduced economic growth and profits is ever present. The EEB therefore tries to show that supporting the repairs sector means new jobs as well as massive savings in raw materials and a reduced burden on the environment. These may be uninspiring considerations from the point of view of politicians, who are judged on the here and now, but if we acknowledge that the fate of future generations is at stake, then it's clear that saving resources and caring for the planet should take priority. So, while the current producer responsibility charges only amount to a contribution towards recycling, the EEB would like to see them extended to preparing electronic equipment for reuse. It is also campaigning for a points system like that employed by ifixit.com, which would tell consumers how easily a particular device can be repaired.

*

Working alongside the EEB is RREUSE, a federation of nearly thirty companies in twenty-four EU member states that deal with the reuse, repair and recycling of electronics, as well as collecting used clothes, furniture, books and toys. The objective is to create jobs, combat social exclusion by putting cheaper second-hand goods back on the market, and to counter the way consumer culture encourages us to discard items long before their time. RREUSE partners employ almost one hundred and forty thousand people. They've rescued a million tonnes of waste from landfill, of which one third was still fit for use. Every year they generate one and a half billion euros in turnover.[121]

In May 2019, RREUSE organised a small conference at Warsaw Polytechnic. The room was barely half full, and most attendees were either employees in the e-waste industry or academics, with a few individuals interested in zero waste thrown into the mix. Among them were representatives of the only Polish RREUSE partner – a company called Ekon, which primarily employs people with disabilities. In their facility, e-waste is separated into twenty-eight fractions, and there are workstations devoted to checking if equipment is still viable and repairing it where possible. The company faces significant financial difficulties, which could be partly resolved if they were to start selling used items. But the system's chronic underfunding means that it's difficult for operators like them to manage their growing logistical and legal obligations: recycling companies and waste processors are required to carry out audits, invest in infrastructure, install monitoring networks and meet strict fire safety regulations.

Mirosław Baściuk from Asekol, a Czech company active in Poland, described how in his country, producers' payments are much higher, leading to a better-funded collection system, which in turn boosts competition, as recycling companies vie for available waste. Asekol has more than three and a half thousand containers for collecting small equipment on Czech streets, financed by manufacturers. This allows them to collect 8.5 thousand tonnes of e-waste every year.

Jerzy Cependa, an electronics handyman known as RepairMan, gave me an insight into product obsolescence from the perspective of someone whose job is to fix things. It turns out we live in the same area. I'd brought a large fan discarded by one of our neighbours to the conference, and Jerzy, originally from Lviv, deemed it worth saving. Over coffee, in beautiful, measured Polish, he explained his philosophy to me. In his eyes, unthinkingly buying new devices, without attempting to fix the old, is as bad as fly-tipping in a forest. After graduating from Lviv Polytechnic with an engineering degree in printing process automation, he came to Warsaw in 2010. In Ukraine, he had struggled to find work in his profession, while in Poland the struggle was to be paid anywhere near as much as his Polish counterparts. It's hard to make ends meet in Warsaw on the 1200 złoty (c. £230) per month he was paid by one large board-games manufacturer.

For several years now, Jerzy has made a living repairing home appliances. Every day he receives a number of requests for help. Sometimes it's as simple as cleaning a dirty gasket or replacing the filter from a washing machine because it's full of dog hair and loose change. Not many people know how to repair things nowadays. We've grown accustomed to the idea that it isn't worth it. But that's not the whole story. Jerzy comes across examples of planned obsolescence in appliances on an almost daily basis. Recently he was hired to fix a three-year-old washing machine that had started making a racket and careering around the bathroom. He found that the stainless-steel drum was made of particularly shoddy material; after three years it had gone from cylindrical to egg-shaped. What's more, the drum had started rubbing through the machine's plastic casing. In other words, it was designed in such a way as to make repairs completely unviable.

White goods are short-lived but cheap enough to be accessible to all. Why bother saving up for more expensive appliances these days – we've grown used to immediately getting what we want. The whole system of micro-loans and interest-free credit encourages the purchase of new appliances. No thought is given to vanishing raw materials, the energy costs of production or the pollution generated by e-waste. And it's not only cheap appliances that aren't built to last. Top-of-the-range fridges, worth many hundreds of pounds and made of decent materials, have dozens of pointless functions that no-one ever uses. When one of the many sensors breaks down, it's often necessary to replace substantial and expensive components wholesale. For decades, a washing machine was a round drum that spun clothes in water – why add a colour screen, reminders, delayed starts and numerous other functions?

Jerzy points out that nowadays every last fitting in an apartment is extraordinarily cheap. Handles, windows, doors, hinges and runners are made of substandard materials. Taps generally can't be reused. Metal parts have been replaced with plastic ones. If you want to fix a leaky toilet flush you can't simply change the seal, you have to replace half the component parts. Each such excessive repair generates waste – and mostly plastic waste at that.

*

While writing this chapter, I was waging a constant battle against my crashing computer. Whenever I left it idle for a moment, it would switch itself off and erase several hours of work, deleting everything I'd managed to write since sitting down that morning. I wrote the part about the dump in Ghana three times. The Blue Screen of Death: a system error that can mean absolutely anything. In the computer repair shop I watched successive clients being sent away with their tickets: "Sir/Madam, this isn't worth repairing"; "This system is no longer updated"; "The parts aren't available". The technician established without a shadow of a doubt that the problem was a broken cooling system. He replaced the fan for 300 złoty (around 50 pounds), but the Blue Screen returned to haunt me nonetheless. I struggled to resist the temptation to just buy a new, problem-free computer, saving my work on a pen drive every ten minutes.

Meanwhile, the oldest working lightbulb in the world is still going strong at a fire station in Livermore, California. It's an old, feeble lamp that has cast a warm yellowish light since

1901. It has shone almost uninterrupted since the year blood groups were discovered and the first Nobel Prizes were awarded. This record-breaking bulb was produced by the Shelby Electric Company. The filament is made of carbon fibre to a design by Adolphe A. Chaillet. As I write, I'm watching the images from an online webcam[122] that sends an updated photo of the bulb every thirty seconds. It's five a.m. in Livermore, there's no-one about yet. The firemen sleep, the old-timer shines on.

Jerzy Cependa called yesterday. He managed to repair the fan – there was a build-up of dust on the mechanism.

The Seat Ibiza is the work of style superstar Giorgetto Giugiaro, considered the foremost automobile designer of the twentieth century. He's the man behind the immortal Golf I, the Maserati Ghibli and many Nikon cameras. What I'm looking at right now, however, are certain unremarkable fragments of an Ibiza's interior. The pillars – which would normally be found support-ing the windows – are made of a mixture of polypropylene and talc, which ensures dimensional stability and prevents plastic shrinkage. There's also a bumper made of polypropylene (PP) blended with the interesting terpolymer EPDM (Ethylene Propylene Diene Monomer rubber), which is resistant to shocks even at low temperatures. Lying next to the pillars I find venti-lators, compartments and various other items torn from the cockpit – the mortal remains of a 1998 Seat Ibiza II. Each part is marked with a serial number, the type of plastic used and the manufacturer's name and logo.

Ever more plastic is used in the construction of vehicles. Metal is replaced by bespoke polyamides reinforced with glass or carbon fibre. In this domain, plastics essentially have no

downside: their low mass offers reduced energy consumption and increased efficiency, while they remain highly resistant to impact. Let's acknowledge that. Imagine the specifications for a metal or wooden car. What would happen if two such vehicles crashed? No wonder the primacy of plastics in the automotive industry is an undisputed fact.

I'm standing at the end of a rough dirt track. The weirdest rubbish finishes up here: a chipped dinnerware set, chairs so broken they look like they've come from the set of a martial arts film, an old sofa bed. I came here for the nature, not the waste, and yet I barely notice the noisy flock of feasting siskins in the alders. I forget to keep an eye out for the great grey shrike, who usually surveys the area from the top of a hawthorn tree.

I cross the meadow and the frost-stiffened grass crunches beneath me, though I also hear the occasional crack of a plastic bottle underfoot.

To the north and west, the land has been encircled by National Road 2, which thunders all day long. Housing estates encroach from the south. Construction waste creeps in from this side too: dug-up foundations, bricks, foam sealant, polystyrene. But there's also furniture from old, bulldozed houses – cupboards, toilets, mattresses, threadbare sofa beds. The shameful secrets of this place are most visible in early spring, when the rubbish stands out among the russet-coloured stalks of goldenrod.

Informing the police when people fly-tip or burn waste is frowned upon, as if we were living by the rules of a prison yard,

willing participants in a conspiracy of silence. The police don't readily investigate such cases, of course, so the conviction rate is close to zero. Who's supposed to clean it up? Local councils, in theory, but it's a fool's errand – new rubbish crops up on an almost daily basis.

It's heartbreaking. The Zakole Wawerskie wetlands are one of the jewels in Warsaw's crown – it's a pity that complicated matters of land ownership reportedly make it impossible to enshrine it as the nature reserve it deserves to be. This Eldorado of the natural world boasts an extensive meadow, which turns into a jungle of tansies, as well as lower-lying land full of reeds, sedges, fluffy willow shoots, a magnificent alder carr, and even an old orchard and an avenue of dying poplars. Undisturbed space, the open horizon and genuine marshland less than ten kilometres from the Palace of Culture. Waste in a place like this is a crime.

7

STILL LIFE WITH POLYMER

By today's standards, the room is practically empty. Just three chairs and a chest, pictures on the walls (some landscapes and a still life) and a mirror that reflects the subject's face. The protagonist of *Room in a Dutch House* leans over her broom, her back to the viewer. We can't tell if she's deep in thought or simply absorbed by her work. Is she humming to herself? Is she silent? On the right-hand side there's an open door, through which we can make out gleaming plates on a mantelpiece. There's an atmosphere of calm – and "Calm" could equally well be the name of this painting by Pieter Janssens Elinga, one of the lesser lights of the Dutch Golden Age.

Their art stands apart. While the Velázquezes of this world depict royal courts, and the Poussins portray muses and flights from Egypt, Elinga gives us an old woman sweeping the floor. But it isn't psychological artwork. This is no Rembrandt, where every wrinkle attests to the depicted characters' inner complexity. The domestic scene is merely a pretext for presenting the stunning play of light on the patterned floor and walls. No metaphysics, because Elinga is neither a moralist nor a scholar of human nature. The Dutch painters believed in artistic freedom.

The broom in art may of course signify cleanliness, and tidying might represent the daily struggle against sin. The mirror could symbolise hubris. But for Elinga, a mirror is just a mirror. Lucky for us. Art in which not every object is an allegory goes some way towards soothing one's intellectual complexes.

And what about dirt? What about rubbish? All gone, swept away. The great artists didn't paint what had been cast to the floor or left to rot in the bins of wealthy traders. It was still too early for that kind of radical confrontation with reality. Rubbish is the invisible inverse, the dark side and great adversary of the bright empty spaces of bourgeois homes.

In English, there's an old expression, "Dutch clean". There was nowhere cleaner than a Dutch home, especially in the stinking, unwashed Europe of the seventeenth century. Whether or not the broom is an allegory, it tells us that dirt and waste – even if they're not seen in the picture – are part of our reality, both physical and moral. That's why brooms feature so often in Dutch art. You can find one in the corner in paintings by Vermeer, by Caspar Nethscher, the favourite pupil of Terboch, and many others. And in the tavern where Jan Steen's *Drunken Couple* are making merry, there are scraps of waste at the feet of the sleeping woman and her foul, bellowing husband. Rubbish as a symbol of chaos and decline, but at least, down there on the floor, it's finally in the right place.

Dutch painters extolled bourgeois virtues: thrift, resourcefulness, tidiness. There was nothing cataclysmic about the garbage of their day. The waste apocalypse wouldn't arrive until the late twentieth century, ushered in by the post-war boom in consumption, single-use items and mass production. Perhaps that's why rubbish has never occupied its rightful place in art and culture. So common, so universal, so timeless, and yet always on the periphery. This dark, dirty, physiological aspect of our reality was avoided in art. Artists – who are people too, after all – averted their eyes from waste just like everyone else. Only in recent decades have they started to offer ever bolder criticisms of consumerism, its costs and the mindless accumulation it entails. The last few years in particular have seen marked changes in the art world's attitude to waste, as pollution and environmental degradation have drawn more and more public attention.

*

After the Dutch there was nothing, for a long, long time. The watershed we're interested in had to wait for a period of unrest, chaos and uncertainty. An era marked by the spine-chilling sense that the world was shaking in its foundations and the centuries-old established order was about to collapse. And no, I don't mean today.

The dawn of the twentieth century brought revolutionary turmoil, angry masses and the disintegration of old empires. The swift tide of history swept up everything in its path, including culture. European art was probably the slowest to respond, reluctant to turn its back on the great edifice it had constructed, on the Sistine Chapel, on Caravaggio's *chiaroscuro* and Goya's

"Caprichos" and to look forwards, towards things that had never before been deemed worthy of an audience's attention.

The biggest challenge was spotting the potential in mass-produced and functional items. Who was first? Was it Braque and Picasso with their *papier collé* – thin pieces of paper and material, sometimes ordinary wallpaper, pasted onto canvas? Or perhaps Vladimir Tatlin, with his counter-reliefs – constructions of wood, plywood, metal sheets and string displayed on the wall? Or Marcel Duchamp, presenting his upturned urinal under the perverse title "Fountain". This was one of the first and most famous *readymades*: prosaic, everyday items that suddenly become works of art by the mere fact of their being exhibited. After all, these objects aren't beautiful in any sense – they're only striking because they're so different from what gallery audiences are accustomed to seeing.

I think the turning point we're interested in was started by a bunch of radicals. Pacifists and anarchists. Dadaists. Enemies of classification, rigid structures and the established order. The real watershed meant a levelling, a reset, a regression to a child-like level of sensitivity, a child whose first words sound like "dada".

We can consider the patron of waste art to be Kurt Schwitters (1887–1948), a follower of the movement but a free spirit, who uttered the immortal words, "Everything an artist spits out is art", a statement that opened a completely new chapter in art and art theory. There had never been and there never would be a more radical credo.

"When he was not writing poetry, he was pasting up collages.

When he was not pasting, he was building his column,[123] he washed his feet in the same water as his guinea pigs, warming his paste-pot in the bed, feeding the tortoise in the rarely used bathtub, declaiming, drawing, printing, cutting up magazines, entertaining his friends, publishing *Merz*, writing letters, loving [...]."[124] We'll leave it at that because I think it's clear enough just how great an eccentric we have here.

Schwitters wanted to make art out of anything he found, be that train tickets, driftwood or bicycle spokes. He collected cheese wrappers, cigar bands, buttons, scraps of cloth, feathers and blades of grass. He believed they could be raw materials just like paint. But while Schwitters' work made use of the contents of dustbins, it didn't yet take a critical approach, it wasn't an indictment of society. The realisation that waste was the leftovers at the end of the long chain of consumption, its unavoidable and unfortunate consequence, would be the next logical step in the progression of art.

*

After the war, Arman Fernandez, a French conceptual artist, took up the baton. His first pieces, "Cachets" ("Stamps"), were printed on cloth and paper and inspired by Schwitters' work. They explored the theme of mass production. Can any metaphor convey it better than stamps as alike as two drops of water, crowded together or scattered across the page at varying intervals?

As far as we're concerned, Arman's most interesting works come slightly later, with his series of "Déchets Bourgeois" ("Bourgeois Rubbish"), that is, the contents of bins enclosed in perspex cases. The artist saw their remarkable beauty, their colours and diversity. He didn't arrange, style or coif the waste. He threw it into the cases and let it create its own composition. Wrapping paper, cigarette packets, cardboard tea boxes, old clothes, egg shells, sardine tins. It's always interesting to observe the universe of a stranger's bin; after all, it gives us an insight into the most private, intimate parts of someone else's life.

In 1960, Arman crammed the Iris Clert Gallery with tightly packed rubbish that overflowed onto the street. The exhibition was called "Le Plein" ("Fullness") and it was a response to an exhibition by Yves Klein, "Le Vide" ("Emptiness"), held there a few years earlier. Klein had presented a striking, empty, white-painted space with a blank window display. That void was then filled with furniture, scrap metal, cardboard boxes and ordinary garbage. An idealised, unreal, aseptic space and a pointedly prosaic, dirty, chaotic jumble of items. And so once again we have overwhelming excess, bringing to mind Calvino's vision of Leonia.

That same year saw the launch of the Nouveau Réalisme movement, inspired by the Dadaist break with form. According to its founding legend, three friends decided to become masters of the world, dividing it among themselves. Yves Klein took all things living or organic, Claude Pascal was to rule over anything natural but not alive, such as rocks and stones, while Arman had dominion over man-made items. He was a hoarder by nature, perhaps taking after his grandmother, who kept drawers full of wine-bottle corks. Over time he amassed one of the largest collections of African art in the world. He also collected Bakelite radios, watches, Japanese armour and European guns. He exhibited unnerving assortments of items – including gas masks, another echo of Duchamps' *readymades*, this time inspired by the Cold War.

In the late 1960s he moved to New York, not the first to be captivated by the city's energy. The United States – the great accumulation, a world in which the overproduction and ubiquity of mass-produced items wasn't just artistic intuition but an undeniable reality. He created "Poubelle de Warhol" ("Warhol's Bin") in his old style, a classic assembly of rubbish in a display case, but his work also took on American proportions. Perhaps the most powerful piece from his later creative period – in terms of its impact but also its dimensions – is "Long Term Parking", sixty cars embedded in a huge concrete column.[125]

*

Readymades and bourgeois bins caused understandable confusion. Modern art often requires footnotes. It can be disconcerting and difficult to decipher. No wonder the uninitiated have trouble understanding it; after all, art stopped being about universal beauty a long time ago. Art is supposed to "trigger something", but what does that really mean? Creative scandals or intelligent provocations are rare, and what we get more often are cynical, calculated stunts designed to grab attention and promote the artist's name. As a simple, uncouth observer, I struggle with art I'm meant to appreciate "just because". Just because that's what the experts say – critics and curators who communicate in a language I don't understand. That's why I feel a wicked satisfaction every time artistic vision collides with popular common sense.

In 2001, Damien Hirst, probably the most famous contemporary British artist, exhibited an installation at the Eyestorm Gallery consisting of the leftovers from a party: half-empty cups of coffee, beer bottles, overflowing ashtrays, magazines, painting equipment. The work was given a six-figure valuation. But not everyone was so enraptured. The next morning, after each item had been painstakingly arranged in the gallery, a cleaner, Mr Emmanuel Asare, simply threw it all into the bin. He later explained, with disarming sincerity, "As soon as I clapped eyes on it I sighed because there was so much mess. I didn't think for a second that it was a work of art – it didn't look much like art to me. So I cleared it all into binbags and dumped it."

We know the name of the man who accidentally destroyed Hirst's work because for many people he became a hero who, by putting rubbish where it belonged, had declared, "The emperor has no clothes". This isn't art, it's a dupe. But nobody in the art world dared admit it. Panic-stricken gallery workers recreated the installation using photographs, and to be on the safe side, set up a sign saying: "Do not touch". Apparently, the cleaner kept his job and the artist himself described the event as very funny, even if for many observers it reinforced the view that Hirst is a mere charlatan who perfectly understands the modern art market and uses cynical, flamboyant tricks to win publicity and pump up the value of his work.

An almost identical story emerged from the Museion in Bolzano, where two Italian artists, Sara Goldschmied and Eleonora Chiari, exhibited a work entitled: "Where shall we go dancing tonight?". It was a room strewn with champagne bottles, cigarette butts, confetti and scattered clothes. The artists claimed that the work reflected the hedonism, consumerism and scandals of 1980s Italy. How do curators sleep at night, calling something so derivative "radical" or "subversive" art? This piece was also swept into binbags by an oblivious cleaner. The museum director, Letizia Ragaglia, was furious. She explained that it was all the fault of a new employee and insisted she had made it very clear he was only to clean the foyer![126]

Many artists have had similar experiences, so let's list only the really important ones. The first was probably Joseph Beuys, whose ink stain was mopped from the floor of the Düsseldorf Academy of Fine Arts. Martin Kippenberg's "When it starts dripping from the ceiling", exhibited at the Ostwall Museum in Dortmund, was treated somewhat differently. Kippenberg had constructed a tower of wooden slats above a rubber trough, which had a layer of paint on the inside walls that was meant to resemble dried rainfall. The insurer valued it at eight hundred thousand euros. The cleaner gave the trough a thorough scrub; arguably her only sin was in being too conscientious – like any housekeeper worth her salt, she tried to make this everyday object look as good as new. I don't know if she lost her job.

*

The Dąbrowa basin in southern Poland. The trees here are pioneering birches that grow even amid ruins, the lakes are concrete reservoirs, the hills are mining slag heaps. As a child, Diana Lelonek didn't think nature could look any different from the post-industrial Silesian landscapes she knew so well. Few environments are more marked by human activity than former mining sites. Only undemanding ruderal plants take root in the degraded, barren land: common knotgrass creeps through the cracked pavement slabs, horseweed sprouts from the crevices of the asphalt in the empty squares, a fragile aspen trembles in a gutter.

I'm ashamed to admit that I only heard of Diana Lelonek shortly before she was nominated for the Paszport Polityka award. It was Paweł Żukowski, a visual artist, photographer and collector, who told me about her work. Paweł's own artworks include icons burned into wooden chopping boards, inspired by Jerzy Nowosielski, and slogans carved into the varnish on old cabinets. He makes obsessive use of discarded materials. The props for his photography shoots include ordinary cardboard boxes, broken furniture, abandoned house plants. He is probably most famous for his highly publicised campaign outside the headquarters of *Gazeta Polska*, "I am LGBT", a protest at the newspaper's decision to distribute "LGBT-free zone" stickers. It was Paweł who, in his cluttered studio apartment, showed me that waste and the clash between civilisation and nature are key themes of Diana Lelonek's work.

Her first highly acclaimed project, "Yesterday I met the

really wild man" from 2015, is a series of photographs showing a group of naturists immersed in seemingly natural landscapes that betray hints of the modern world. In one of the pictures, they are passing through waste ground, half hidden in vegetation typical of post-industrial terrain. In another, they are climbing up a large heap of sand, a radio mast looming above them. This is nature branded irreversibly by man, anthropogenic landscapes in which only the least demanding organisms will prevail.

The photographs are on a grand scale, wide shots that restore things to their correct proportions: man is small and defenceless set against a backdrop of sky, trees and the vast expanse. Standing motionless amid the goldenrod, the naturists look like deer, like wild animals, just one of the many species surviving in an altered environment.

Lelonek's later works present not so much a premonition of decline as the certainty of impending catastrophe. And an accusation. The only consolation comes from the knowledge that nature will survive, that the world will go on without us. This sentiment pervades the work entitled "Ministry of Environment overgrown by Central European mixed forest", created for a billboard campaign organised by the collective Sputnik Photos as a protest against logging in the primeval Białowieża forest. This photomontage showing a government building taken over by vegetation looks like a scene from a post-apocalyptic film, or a perfectly authentic snapshot from Pripyat, the abandoned ghost town a few kilometres from the

Chernobyl nuclear power plant. The lesson from the Chernobyl Exclusion Zone is that nature will find a way, adapting to new conditions, and that specialised species will reclaim everything we have created.

Lelonek's subsequent works are even more political, denouncing exploitation, overproduction and the capitalist pursuit of economic growth, which is slowly pushing us over the precipice. Man is guilty, and that guilt has roots as ancient as civilisation itself. In "Zoe-therapy", another attempt to redress the disrupted balance of our relationships with the natural world, Lelonek exacts a playful revenge on the founding fathers of European thought by allowing fungi to sprout over portraits taken from Władysław Tatarkiewicz's *History of Philosophy*. Great thinkers trapped in Petridishes. Darwin may be Darwin, but regardless of his accomplishments, *Aspergillus versicolor* will still thrive happily on his face, so long as the conditions are favourable.

Waste takes centre stage in the project "New Archaeology for Liban and Płaszów". Here, Lelonek seems to be playing with the conventions of archaeological articles. She presents photographs and objects from the grounds of Płaszów concentration camp and the neighbouring Liban quarry in Kraków. There's a broad range of finds: ordinary consumer waste, packaging, the casing from a hairdryer, bits of an Apple computer charger and an old PWN Encyclopaedia, open to the letter E and its entry on Europe, its creased, damp pages resembling the gills of a trampled toadstool. This former quarry and killing ground is now the site of an illegal dump overtaken by wild, indifferent nature, home to certain abandoned elements of the set of Steven Spielberg's *Schindler's List*, which was filmed there. Jewish gravestones and camp fenceposts in a place where they never stood, hinting at false historical depths. Today, unsuspecting tourists take them to be shocking reminders of the war.

*

Waste is a convenient metaphor for everything. It speaks of excess, chaos and the fleeting nature of our world. The internet as a dustbin. Junk food, junk mail. Soon we'll start talking about junk lives. No wonder, then, that the waste apocalypse is a fashionable topic, and is growing ever more so. In this sense, Diana Lelonek has artistic relatives scattered across the world.

"In March 2008, I found out that there was a 'floating land-fill', about the size of the United States," writes Maarten Vanden Eynde on his website. "[. . .] Almost nobody knew about it at that time so I wanted to raise awareness about this incredible phenomenon and find out what could be done with this new 'raw' material. In January 2009 I visited Charles Moore, marine researcher at the Algalita Marina Research Foundation in Long Beach, who discovered the Plastic Garbage Patch in 1997. He gave me a first sample of plastic debris from the North Pacific Gyre which I melted into a small plastic coral reef, the size of a football. The trash became beautiful again and seemed to solve two problems at the same time: the plastic in the ocean and the disappearing coral reefs around the world.

"I decided to make the 'Plastic Reef' as big as possible and went to the Hawaiian Islands, which are located in the centre of the North Pacific Gyre and are getting an incredible amount of plastic flotsam on their beaches. [. . .] In February 2010 I joined the Pangaea Explorations on their boat *Sea Dragon*, which is doing research on plastic pollution worldwide. We crossed the Atlantic Ocean to gather as much plastic as possible and melt it into the growing 'Plastic Reef'."[127] The artist also travelled to other gyres in the Pacific, Atlantic and Indian oceans. For another piece, entitled "1000 miles away from home", the artist placed some of his finds inside five snow globes. When you shake them, the plastic particles swirl around as if in the ocean.

Vanden Eynde is fascinated by our legacy: how long will man-made items survive and what will our civilisation leave behind? Plastic is a material that in optimal conditions will never decompose. It is considered to have already left a significant, lasting mark on the geological structure of the Earth. "Plastic Reef" is a bitter and pessimistic project, but Vanden Eynde also approaches these issues in a more playful way. At the news that the IKEA catalogue had broken the Bible's record as the most prolifically printed and distributed book in history, Vanden Eynde scaled the fence of the Forum in Rome and buried one of the store's teacups in an alcove. It might not be a particularly original object to leave as a souvenir for later generations or future explorers, but it is significant. It will tell them a lot about the civilisation of our time.

"Plastic Reef" is an interesting counterpoint to the mysterious phenomenon of plastiglomerates. In 2006, Charles Moore, who has been mentioned several times already in this book, found some strange stones formed out of bits of plastic, sand, shells, pebbles, ballast, coral reefs and wood washed up on Kamilo Beach, a popular surfing spot in Hawaii. He photographed these enigmatic hybrids but gave them little further thought. Indeed, plastiglomerates might have waited a long time to be discovered anew, if Moore hadn't shared the picture during a lecture at Ontario's Western University in 2012. Patricia Corcoran, a researcher, was intrigued by these strange combinations of natural and synthetic materials, whose origin was unknown.

Her team collected twenty-five objects from Kamilo Beach, ranging in size from "a peach pit to . . . a large pizza". The objects were formally defined as a type of stone, a fusion of rocks and anthropogenic waste. They were found to contain fragments of sailing ropes, fishing gear, plastic pellets and various tubes, pipes, lids and caps. The most convincing explanation holds that the elements were fused together in beachside campfires.[128]

In 2019, Kelly Jazvac exhibited a plastiglomerate as a *ready-made* in a gallery.[129]

*

Once discarded, waste takes on a new role that is completely different to its original function. In favourable conditions it quickly becomes cloaked in plants, fungi and lichen, blending into the natural landscape. Every piece of litter can be used, and every piece can be useful. *Center for the Living Things*,

Diana Lelonek's most ambitious initiative, is a type of scientific–artistic institute that gathers waste items that have become breeding grounds for natural tissues, such as fungi, mosses, vascular plants – in short, anything that can thrive on the opportunities provided by the things we discard.

Lelonek's underlying premise is that untouched nature does not actually exist. She calls on the scientific community to accept the terminology of "waste-plant habitats" and to recognise the pervasive anthropogenic influence on the environment, ambitiously encroaching into the domain of science, into disciplines reserved for botanists, mycologists or bryologists. And it's true that the distinction between the natural and the man-made has been fading before our eyes for years. Birds live in city street lamps just as they used to live in forest hollows, and mountain lichens cover the Palace of Culture just as they do rocky cliffsides. Many plants now seem to have outgrown patterns of behaviour that have been observed by botanists for centuries.

Anyone who has collected litter knows how illegal dumping grounds often have their own specialisms. In the forests near Łódź you can find heaps of textile cuttings, while on the outskirts of the city it will be dumped rubble from housing developments. Diana Lelonek proposes that descriptions of these sites should indicate the type of waste in question. So, a place where a clothing factory discarded its materials might be called a "textile habitat", and a ditch filled with debris from a DIY renovation could be a "post-construction habitat". In

Lelonek's proposed classification, it is also important to reference the original characteristics of the environment where the waste has been discarded. This results in exotic-sounding hybrids: a "post-construction meadow-type habitat", for instance, or a "polyurethane forest-type habitat".

Art lovers and gallery visitors may be taken aback by the unusual hybrid structures that arise from intertwining plants and polymer fibres: moss-covered carpets or upholstery; a shoe with a wild strawberry growing out of it. But these waste discoveries are equally fascinating for scientists. Consider, for example, a polyurethane sponge found outside Warsaw, which had spent years exposed to the elements. As with any decomposing material, its surface had become more porous; a layer of organic matter had formed in the hollows. Foam also stores water very well, and thus a bryophytes' paradise was formed. The rough surface, which incidentally looked a bit like rotting wood, provided easy, welcoming grips for the rhizoids of these mossy plants. Nappies act like sponges in that they also preserve moisture, providing a home for star moss – an exceptionally resilient moss that can spend years waiting in a herbarium for its moment to come.

It seems that for many organisms, man-made waste is a base like any other. Nature isn't picky. One such habitat was a piece of polystyrene foam found by the Vistula, close to Grudziądz, thoroughly mixed into the natural soil. It was a "Styrofoam stone", overgrown with a moss, *Leskea polycarpa*. This moss prefers environments that are periodically inundated, like thickets, roots or riverside stones, so you could say it treated the polystyrene like a rock. Bryophytes' rhizoids are even capable of attaching themselves to the apparently smooth surface of a PET bottle. They may actually accelerate the natural decaying process, squeezing themselves into small crevices that are invisible to the naked eye.

Nonetheless, the undiscriminating organisms that colonise waste habitats can be sensitive to changing conditions. They don't take well to controlled environments, and they aren't easy to cultivate, as if they want to decide for themselves how they live, grow and die. Every time the exhibits are moved, there's a risk of damaging the delicate root structures and cutting the organisms off from life-giving nutrients and moisture. Bringing them inside galleries or air-conditioned rooms can dry the plant litter out, with fatal consequences.

The exhibits in *Center for the Living Things* are watered by a sophisticated system of pipes that reacts to changes in the surroundings. This allows life in the cabinets to continue. But there are other threats: sometimes the dense substratum shelters other lichens, mosses and plants, or even insect eggs, which in the right conditions may hatch larvae. One piece fell victim to a plague of woodlice, which decomposed the organic matter. The warm, moist environment is also an ideal habitat for mould, so this seemingly self-sufficient collection requires close monitoring.

Some time ago a Belgian collector approached Lelonek, wanting to buy one of the exhibits. But he had no interest in ensuring the wellbeing of the organisms living inside it, so the artist refused. *Center for the Living Things* means life, constant change, a process. It can't be cut short.

Many of the specimens are kept in the Botanic Gardens in Poznań, in a disused building and broken greenhouse that once belonged to a research group. It's a brilliant, post-apocalyptic setting for the whole collection. But surprises still happen – seeds carried in on the wind led to an invasion of evening primrose, which colonised the greenhouse. Lelonek hopes that the waste-plants and their chosen footholds will become permanent features of the gardens' displays. Given that they already house plants that grow in coastal sand dunes, why shouldn't they showcase the world's first exhibition of organisms that have made their homes in polymer, footwear or construction habitats? Why not present plants that flourish on post-mining slag heaps or in degraded post-industrial landscapes? They may not be traditional environments, but in today's world they are commonplace and impossible to ignore.

*

About a decade ago, New York's Guggenheim Museum exhibited two works by the Mexican artist Gabriel Oroczo: "Astroturf Constellation" and "Sandstars". The former is an assortment of ordinary fragments of rubbish that we come across every day. They were all found on a sports field in Manhattan. Twelve hundred items are displayed inside a glass case measuring 150 centimetres in length. They include torn shoe soles, remnants of balls, chewing gum, sweet wrappers, screws, buttons, coins.

They're scraps – so small we don't even recognise them as litter that ought to be removed.

Oroczo's second work, "Sandstars", presents a similar number of items found on beaches around Isla Arena, Mexico, a wildlife reserve that is simultaneously a whale mating ground and their ancient cemetery. Few people have access to the site, meaning that waste washed up by the sea is essentially left undisturbed. Oroczo's items are displayed in a rectangle, and include fishing nets, ropes, ships' headlights, dozens of bottles of varying shapes, sizes and colours, crumpled, faded buoys, mooring lines, helmets, tennis balls, lightbulbs and even hardened rolls of toilet paper.

"Sandstars" is mesmerising. A constellation of shapes and colours. The exhibit's arrangement and style bring out the individual beauty of these prosaic objects. As the *New York Times* wrote,[130] there's something alchemical about Oroczo's work – the chaos of discarded litter is turned into art. Rubbish placed in a museum becomes art: stripped of utility or value, it becomes priceless. But more than anything else, "Sandstars" shows the apocalyptic scale of the unfolding environmental catastrophe. It's especially painful in Isla Arena, a nature reserve supposedly protected from human interference. As is often said, waste knows no borders. No place on earth is spared.

Last winter, my friends Zosia and Łukasz sent me the catalogue for Tadashi Kawamata's installation "Over Flow", exhibited at the MAAT gallery in Porto. The artwork makes reference to the famous Japanese print "The Great Wave off

Kanagawa" by Hokusai – only Kawamata's wave is made of waste. For years, his work has focused on the sea and marine litter, and "Over Flow" consists of items washed up on beaches and collected by volunteers. They are strewn over nets that stretch across the gallery space, creating an impression of polluted surface water. You can descend the gallery staircase to see it from below, where the suspended waste creates an oppressive, claustrophobic sensation. These two dimensions are connected by a sunken boat lying on the floor. Kawamata's recreation of the sea's depths offers us an equally valid, though usually inaccessible perspective: rubbish on beaches is barely a fraction of what is swarming in the sea, moving in all possible directions.

*

Illegal dumps. They're usually located some distance from built-up areas, by dirt roads, on the outskirts of cities. Diana Lelonek sometimes struggles to transport her future exhibits. Occasionally she's had to take the waste on trains and in taxis. It's best not to reveal what's inside the package, people are unlikely to be all that understanding when they find out it contains moss-covered rubbish. But Gwarków Street in the Targówek district of Warsaw always has heaps of waste to offer her. Homeless people burn cables to recover metal, and building-repair companies and cleaning firms have dumped their refuse here for years. We've walked barely two hundred metres and we're already dragging a huge piece of car upholstery behind us.

At the foot of an old plum tree, we find glass pill bottles, old tubes, aerosols. This is where a pharmacy disposed of its waste. A little further on, there's a carpet, three balls, a splintered cupboard, rollerblades, some roof tiles. A pile of plastic guttering. A piece of old, greyed glass wool, overgrown by moss. There's a memento from a First Communion. The name of the child and their parish, once recorded in the blank spaces, have succumbed to damp. Next we find a large hardback copy of *The Sanctuary of the Heart of Jesus the Merciful in Kalisz* and the programme for a concert of music by Johann Strauss, entitled "The Waltz King", from its avant-premiere in December 1977. We collect three bags of waste, which Diana will turn into art. To clear this site after years of fly-tipping you'd need to bulldoze half a metre deep into the earth.

A half-litre PET Coca-Cola bottle. It's been lying on the railway line near the Warsaw Water Filters for some months now. I cycled past it all spring and summer. In autumn it seemed to disappear and I thought someone must have picked it up, but no, it had just shifted a few metres. It's becoming harder and harder to spot: the label is fading and the bottle is coated in street dust and rust from the rails and is slowly blending into its surroundings. There's little chance of anyone picking it up, people tend not to come this way. I thought it was a good, safe test case, a place where rubbish lies in plain sight and won't harm anyone.

PET bottles are one of the most common types of litter in our environment. You need only walk down the hard shoulder of any road or glance inside a random bush to find one. They're discarded as easily as they're bought, and there's no way this will change until an easily accessible deposit system is brought in. Bottles are everywhere and they fit perfectly into the dynamic big-city lifestyle that's promoted all around us. After all, an active young person doesn't sit with their coffee in a café, they

drink it from a single-use cup while running for the bus. And it's the same with light, convenient plastic water bottles. We're always being told that water is the secret to good health, well-being, and who knows, maybe even to success.

With the spread of the Western lifestyle in Asia (especially in China), the consumption of water in plastic bottles will rise. It's already estimated that twenty thousand bottles are sold every second around the world.[131] Every second! Four hundred and eighty billion bottles of water were produced in 2016. Less than half were collected and recycled. Barely 7 per cent was turned into recyclate used to produce new bottles. That's a shame, because PET is a very easy plastic to recycle; in Poland it can be treated in SSP plants, that is, through solid-state polymerisation, which does not reduce the quality of the output.

Polyethylene terephthalate has been used massively for almost half a century; the first PET Coca-Cola bottles rolled off the production lines in 1978. It has a particularly wide range of applications in the packaging industry – in trays for packing meat, for example. Objectively speaking, polyethylene terephthalate is a wonderful invention. Light and resilient, it forms an impenetrable barrier against external conditions, and it can be transparent. Not all plastics can do that.

Many scientific articles compare the impact of glass and PET bottles on the environment and human health. Life Cycle Assessments (LCAs) analyse energy, water and resource consumption and study the carbon footprint, soil acidification, eutrophication and effects on the ozone layer caused by the manufacturing process. Some include further factors. Depending on the methodology adopted, either plastic or glass triumphs. Plastic is light, so transporting it generates significantly lower carbon dioxide emissions. It's more shock resistant, too. PET can also be effectively recycled, though not in as many places as container glass.

The real problem is the inefficiency of the current system: massive production that is in no way counterbalanced by recycling. Few companies voluntarily use recyclate (like the Dutch company Barle-Duc, for instance), which is where regulation comes in. By 2025 all PET bottles manufactured in EU countries must be made of at least 25 per cent recycled material. But what kind of bottle would I choose if I forgot my reusable one? A glass one. And it's always a bonus if the water hasn't come from the other side of Europe. I really liked the taste of Vichy Catalan water, but I can live without it in Warsaw.

8

ENVIRONMENTAL NEUROSIS

About two years ago, I self-diagnosed an illness I call environmental neurosis. Like many inhabitants of our planet, I bear a sense of guilt that I'm contributing every day to the extermination of our world. I try to change my lifestyle, my habits, but it all feels futile. Buying used clothes, riding a rusty bicycle, washing laundry at 30°C – it might help me feel a little better, but it isn't going to save the world. Perhaps what bothers me most is precisely this feeling of helplessness. I label my endeavours environmental neurosis, an essentially harmless illness or eccentricity, as a way of distancing myself from my own struggle. All day, every day – in wakefulness and in my dreams – I fight to allay my fears for the future.

7.10 a.m., (sometimes 6.42 a.m., often unfortunately 7.42 a.m.) – I wake, I wash

It begins in the shower. I try to do as they teach – a little water, lather and rinse – though I really enjoy a good soak in the bath. I'm even depressed by the several litres of clean, drinkable water that pour down the drain with each flush of the toilet. Some people are seized by neurosis as soon as they wake. Marta reaches for her phone while she's still in bed, only to stress about the fact that it's consumed 13 per cent of its battery overnight. What on earth was it doing? She's going to have to charge it again soon.

It's winter and the bathroom's cold because I'm trying not to use the heating too much. I stand in the shower, freezing and soapy, and it strikes me that it's quite pointless, actually. Will these few litres of warm water change anything? That's what Polish Water, the government agency, suggests in its "Stop the Drought" advertising campaign. It encourages us to collect rainwater and wash our cars less often.[132] But is the drought really the fault of those who use drinking water to keep their gardens looking nice? Let's keep things in perspective. The campaign failed to mention that over 70 per cent of water in our country is used by industry. Like the energy industry, for instance – which pumps it out of open-pit mines and uses it for cooling power plants and washing coal. The wastewater from this process is a thick suspension, so highly polluted that it can't be reused. It's really not cool when the big guys shift the blame onto the little guys.

That's not all. Mining leads to the formation of cones of depression, which suck in groundwater from far and wide. Dr Sylwester Kraśnicki, a hydrologist and expert in the environmental impact of mining, estimates that the worst-case scenario for the new opencast mine in Złoczewie would see the cone covering 3,100 square kilometres[133] or 1 per cent of the area of Poland. Anyone who's seen the disappearing lakes in the province of Greater Poland has seen what's it like to live near a mine.

Dr Kraśnicki also criticises the planned "Jan Karski" mine on the edge of the Poleski National Park, which will cause the surrounding area to dry out. Both human and non-human inhabitants of these lands will experience reduced access to water. It may lead to the destruction of priceless ecosystems, marshlands and habitats for European pond turtles, among other species. The mining activity may cause rocks to crack and water to drain from Groundwater Reservoir 407 (the main body of groundwater for the Lubelskie province).[134] And, of course, the salted, heavy-metal-polluted water from the shafts will end up in the nearby Wieprza and Mogilnica rivers.

I dry myself and think about other plans for drying out our country. There are shady megalomaniacal plots to restore inland water navigation, which bring to mind the catastrophic effects of similar Soviet engineering projects. Just take the proposed E40 Waterway, linking the Baltic and the Black Sea, part of which is to be a canal. Experts are agreed that all three options for its route currently under discussion are absurd. The rivers that would supply the canals – the Wieprza, Tyśmienica, Bystryca and Wilga – do not hold enough water to support navigability, while pumping it from the Vistula would be very expensive, and diverting it from the Bug river would without doubt result in ecological and economic disaster: according to a report commissioned by the Frankfurt Zoological Society, the frequency of severe droughts would rise by 172 per cent.[135]

I step out of the bath and look at the row of shampoos, conditioners, lotions, body washes and soaps. Many of them come in PET bottles. These products should really be sold in HDPE (high density polyethylene) or PP (polypropylene) containers – PET should be used only for food packaging. When PET is used for chemicals, it can't be recycled in Poland and has to be separated from the waste stream. This seriously complicates the work of sorting sites and makes it difficult to achieve target levels of recycling. Pure absurdity. Why is it allowed?

7.42 a.m. – I stand in front of my wardrobe

Like almost everyone else, I have more clothes than I need. For decades, I've kept a British Rail train driver's coat that my mother gave me. I've never dared wear it. And a pair of shapeless black jeans; tight at the thighs and baggy at the calves. A panic-buy for a funeral. I also have a tweed jacket that only Sean Connery could pull off. Fortunately, most of these pieces are second-hand, but many of us have grown accustomed to regularly buying new items. Fashion changes every season, and clothes from big chains are cheap and tend to be poor quality, so you can get rid of them without feeling guilty.

I put on my black jeans – a different pair, slightly too big, with a tear at the thigh. I ripped them five years ago on a fence surrounding the house of a Prussian ornithologist I was researching. I think I've had them for about a decade, and I have to admit they do no favours for a man who is not as young as he once was. But I can wear them to walk the dog. I hate throwing away clothes, especially when I remember that barely 1 per cent of what's discarded is fully recycled (from clothes into clothes). Another 12 per cent is downcycled – used to make mattress filler or insulation – but almost three quarters end up in landfill or incinerators. And it isn't just old clothes that are burned. Production is so high that companies destroy the stock they aren't able to sell – especially luxury brands, who need to maintain the exclusivity of their products. In 2017, Burberry burned clothes worth 40 million dollars.[136] Cheaper brands have also been accused of such practices. Many firms intentionally destroy unsold clothes so that they can't be re-used. How is such wastefulness even legal?

My wardrobe contains clothes made from cheap industrial cotton and from synthetic fibres. What's worse for the planet? Chluin Kloin pants (5 per cent lycra) from the market, or a T-shirt of unknown materials (the text on the labels was washed away long ago)? Cotton crops take up 2.4 per cent of all fields, but they account for 16 per cent of insecticides and 6 per cent of herbicides used in agriculture.[137] Furthermore, cotton requires vast quantities of water, and this can lead to environmental disaster: the artificial irrigation of fields with water from the rivers that fed the Aral Sea led to its gradual disappearance. On the other hand, synthetic fibres used to make clothes pollute water with microplastics. Polyester, used to make fleeces among other things, is one example. And synthetic additives are found in almost all clothes, as anyone who has tried to find new cotton jeans in a major chain will know. Depending on the material, one kilogram of laundry releases between 640,000 and 1.5 million particles of plastic into the environment.[138]

The relentless pace of fast fashion is horrifying. It's estimated that in the last fifteen years the clothes industry has doubled its turnover.[139] No wonder, since we replace our clothes much more often than we used to. They spend up to one third less time in our wardrobes. Massive overproduction is rife. I saw this for myself two years ago. I was taking my clothes to a homeless shelter and I felt like a hero, glowing from within. I expected gratitude and a pat on the back, but nobody in the room where they were receiving the clothes even glanced my way. Someone pointed to where I could leave my two bags. A few people were attempting to sort through a multistorey heap of tracksuits, jackets, shoes, belts, sweatshirts and jumpers. I wasn't the only one who had the noble idea of getting rid of my unused clothes before Christmas.

Every morning I look at the row of shirts I never wear; the old, stretched woollen jumpers; the fleeces I don't use so I don't have to wash them; the masses of clothes I'm embarrassed by but feel guilty about throwing out; the three pairs of faded trousers that I've been promising myself for ages I'll dye one day. And despite two overflowing wardrobes, I wear the same thing every day – an old pair of jeans and one of the more respectable-looking sweaters.

Now, this definitely isn't the biggest problem facing our civilisation, and many people will find it totally silly and trivial, but I can't help wondering every day, on my morning walk, how to dispose of my dog's poo in the least harmful way for the planet. Using at least two plastic bags per day alongside all those efforts not to accumulate them obviously makes no sense. You can buy so-called biodegradable bags – not the flimsy LDPE kind but rather, judging by the soft, cloth-like texture, something made of starch – but "biodegradable" is a vague, overused word that in no way guarantees that the bag will disappear without a trace like an apple core. People still happily buy them, though, with the warm feeling that they're doing something good for the planet.

Most of those bags, in accordance with standard EN 13432, do biodegrade in composting conditions. But not all will break down in home compost bins, and many of them require constant high temperatures and moisture, which can only be provided by industrial installations. If a bag ends up in landfill, weighed down by tonnes of rubbish in an oxygen-less environment, it won't decompose. Today, in winter 2020, park refuse is still not collected separately. It's treated as mixed waste. No-one is going to mess around composting dog poo in accordance with the bag manufacturer's lengthy instructions. Besides, in Poland, excrement isn't composted anyway. You're not allowed to throw it in the organic fraction. Perhaps the best option is to incinerate the bag, but in that case it doesn't really matter what material it was made from.

The first city to force its residents to clean up after their dogs was New York. It was the result of a long campaign by the indefatigable social activist and media personality, Fran Lee. Lee, a former Broadway actress, was a frequent guest on local television, where she would speak passionately about, among other things, sweeteners, asbestos or "how to make a candle out of sausage".[140] But it was her fight against dog waste that really made her a household name. In 1978 she successfully lobbied for the introduction of article 1310 of the city's Health Code: "It shall be the duty of each [. . .] person having possession, custody or control of a dog to remove any faeces left by his or her dog on any sidewalk, gutter, street or other public area." Similar provisions were later adopted in other American cities.

The poo problem has been solved in different ways. In Paris, starting in 1982, they deployed *motocrottes* – motorbikes with a vacuum cleaner that sucked it into a container. This brainchild of Jacques Chirac, then mayor of Paris (Parisians called them *chiraclettes*), survived twenty years. They disappeared from the streets in 2002 as maintenance costs were considered too high. Now Paris just issues fines to those who don't clean up their mess.[141]

In truth, it was a dreadful idea: I don't understand why owners should be absolved of their obligation to tidy up after their dogs. Every now and then, someone eagerly suggests: "What if we buried it?". A cry that contains all the excitement of a young seaman in the crow's nest who at long last spots dry land. Well, it's easy to predict the consequences – the grass, churned up time and again, would soon become mud. The slow, oxygen-less decay of the excrement would probably present an epidemiological threat, and there's a risk of over-fertilisation, which would kill off the surrounding vegetation. I read that dog waste can be composted in a garden composter,[142] but it's hard to imagine these being installed on every street corner. How would we bring the poo to them? Who would take care of the compost? It has to be mixed, moistened and supplemented with plant remains. Of course, there's the option of not cleaning up after our canine companions, but then daily life in our cities would become even more of an ordeal.

In New York, health agencies have for some time recommended collecting the poo, taking it home and flushing it down the toilet. Perhaps I'm lacking in imagination, but what are you to meant to carry it in? Sometimes I use ripped carrier bags, sometimes paper bags from the bakery that aren't fit for anything else. I only pick it up in town. In forests or fields, I let it be. I don't have a perfect solution. Maybe this one really is an insoluble problem?

8.13 a.m. (in summer and winter)

Every day, there's a car idling in front of my house with its engine running, while the driver sits inside, attentively scrolling on his smartphone. Waiting. In winter he has the heating on, in summer the air conditioning, but the fumes belching out of his exhaust pipe are the same all year round. I ought to pull him up on it, but I'm not a naturally confrontational person. My partner says I'm accommodating; I don't like those emotions, that tightness in the throat. I play through the exchange in my head, thinking how to start a conversation politely but decisively. How to explain it and not get mired in some pointless dispute.

I almost always lose against myself. I turn on my heel, swallow hard and go on my way. I don't want to ruin my day. And then I resent myself for it – I should have said something.

8.15 a.m. – breakfast

I've managed two years of lacto-ovo-veganism with episodes of ovo-veganism and insignificant dalliances with veganism. I gave up fish without much trouble, especially after almost eating the plastic ball in the barracuda. I confess to giving myself a dispensation for my annual visits to Holland; raw herring with onion is basically the only thing Dutch cuisine has to offer me. If someone had told me five years ago that I wouldn't be eating meat, I'd have told them to pull the other one. It used to seem somehow extreme, unnecessary, excessive. But one day I looked a pig in the eye as it was being sent to the abattoir and that was that, from one day to the next.

Today I'm fighting my weakness for cheese, unfortunately with limited success. Yes, I'm aware of the horrors of milk production (I'm a hypocrite and not at all proud of it). When it comes to eggs, I have to trust those stamps on the shells and the rustic scenes on the boxes. My favourites are the ones we buy in a village in Kashubia. The hens there have a big yard behind the barn, they spend all day digging holes, basking in the sun, chatting, hopping, clucking. The owners once put out an old furniture set for them. To me, a hen snoozing in an armchair is the happiest hen there can be. I eat eggs like that with an (almost) clear conscience.

Our daily sustenance is an ethical minefield – and, of course, a pile of rubbish as well. Packaging takes up half the table. After a week, it's spilling out of a large IKEA bag. People are annoyed by the huge amounts of plastic in food shops, especially in the fruit and veg aisles of supermarkets. Nearly 40 per cent of annual plastic production is driven by packaging. Nonetheless, we're still a far cry from the United States, where I've seen marvels like apple quarters sold on a laminated tray. But why bother with plastic when most fruit and vegetables have a peel that seems to protect its contents well enough? The thing is, it may provide some protection, but fruit peel wasn't designed with supermarkets in mind, nor the hundreds of people who poke and prod their wares. Food that goes to waste because it's been handled and chucked around is a real problem.

Once harvested, fruit and vegetables are taken to a ware-house or cooler. There they await the next step – packing and distribution. Packaging is meant above all to shield the food and prolong its shelf life. Research shows that cucumbers wrapped in plastic remain in peak condition three times longer than loose ones. That means less frequent deliveries and thereby a smaller carbon footprint and less waste.[143] It's also often stressed that plastic is lighter and more flexible than glass or cardboard, which further limits emissions. Packaging, say food producers, protects food and prevents waste.

But is it really that essential? Food waste is undoubtedly a massive problem: it was estimated in 2015 that an EU citizen wastes more than 170 kilograms of food each year, so anything that reduces that can't be an entirely bad thing. When it comes down to it, doing away with plastic packaging would require redesigning distribution networks and ultimately the return to a more seasonal diet. We know that food travels around the world, but we generally don't realise the scale of this phenomenon. According to British customs data, in the first half of 2018, the United Kingdom imported 23.5 billion pounds worth of food and drink.[144] No-one is really suggesting we deprive ourselves of the privilege of eating strawberries or grapes in the middle of winter – the only question is when such goods will be taxed appropriately, to take into account the true cost of transporting them.

Packaging may prevent food waste, but in many cases its main purpose is marketing. It's there to grab our attention and force us to buy more of a product than we really need.[145] Instead of two tomatoes we'll buy four, one of which we'll end up wasting. Instead of three onions, we'll buy a kilo in a plastic bag. At home we'll discover that two are already mouldy. And though Life Cycle Assessments favour plastic packaging because of its low mass and the resulting reduced transport emissions, it's worth pointing out that sometimes these analyses don't take into account important factors such as recyclability. The poor recovery rate, low value and considerable environmental impact at the end of their life cycle is the true, hidden cost of plastic packaging.[146]

11.20 a.m. – neurotic procrastination

Bad news is part and parcel of my daily life. Every day, I come across articles about the destruction of priceless nature, the immorality of eating a South American avocado or how tourism is killing the planet. There are even tools for generating distressing news: you can do an online test to calculate your carbon footprint. Of course, all these mathematical models are simplified, so the result is symbolic. "If everyone lived like YOU, global emissions would increase 1.4 times. To stop climate change, these emissions would have to be spread over 7 planets. Unfortunately, we only have one," admonished the calculator on ziemianarozdrozu.pl. It worked out that I produce seven tonnes of carbon dioxide per year. The version on www.footprintcalculator.org was a little kinder about my lifestyle. If everyone lived like me, we'd need 2.2 planets. The results were depressing, but lower than average. That made me feel a little better.

We have no influence over many of the factors that contribute to our daily carbon footprint, especially if we live in a city.

We can't do much about the fact that the heat in our homes comes from a network fuelled by coal, as we don't have the space or means to install solar panels. It isn't always easy to access local, seasonal food. We have no influence on the technology behind sewers, waterworks, street lighting, office buildings or on emissions from hospitals, theatres, museums, sewage treatment plants and schools. Who has the energy to fight over these things? It is how it is – our world is designed to be convenient for us, not safe for the environment.

According to calculations by the Global Footprint Network, we've been living on credit from planet Earth since the 1970s. That means that for nearly fifty years, our exploitation has exceeded the planet's natural capacity to renew resources. You can dispute that fact, argue that the simulations are imperfect and based on inflexible models that leave out many elements, but you cannot fail to hear the trumpet of the apocalypse or see the felled forests, melting glaciers and piles of waste floating in the oceans. It's said that irreversible damage will be done in little more than a decade.[147] That deadline is absurdly tight. We all hope the scientists are wrong. But for many of us, eleven years is long enough to make ourselves at home up shit creek.

We've grown accustomed to convenience, and we won't want to give it up, to abandon all those privileges of civilisation that are so self-evident we no longer notice them. I say "we", but of course convenience isn't the purview of all inhabitants of this planet. I say "we" meaning our Western society, enjoying comforts that are far beyond the reach of most *Homo sapiens*

sapiens, even in their dreams. People on the other side of the world are so far away, so unlike us, that we feel no practical responsibility for their fate. We're equally indifferent to the fate of rare species that are disappearing for ever. Our prosperity is built on want, on the civilisational divide, on the cheap, slave labour of our distant brothers. We live at the expense of the environment in which they live, the air they breathe, and the water they drink. Wealthy societies will be able to absorb the consequences of the catastrophe for many years, but the poor will have nowhere to hide.

2.00 p.m. – a casual lunch in town

Obliviousness irritates me. Why are people so obsessed with disposables? Why do most people in Asian diners insist on fumbling about with those bamboo chopsticks? Are they trying to show how worldly they are? There's always reusable metal cutlery available. Why do people in cafés drink coffee in single-use cups even when they can take a normal mug? They drink, they discard, they leave. Is it really so chic, so stylish, so metro-politan? Are they trying to convince themselves they're in an American sitcom?

3.30 p.m. – bleak thoughts on my second dog walk, an encounter with catkins on pussy willows in January and a sighting of cranes above the city

The snowdrops haven't had to work very hard this year; they blossomed on Dantyszek Street before a measly, one-day snowfall. You can count the number of feeble night frosts this winter on the fingers of one hand. Anyone already suffering from climate anxiety was no doubt joined by millions more after the publication of an IPCC report in 2019. To have any chance of abiding by the Paris Agreement (to limit global temperature increases to 1.5°C), nations need to reduce their carbon dioxide emissions over the next decade by 7.6 per cent each year. So, in 2030 they need to be more than 50 per cent lower than in 2018. But global emissions are actually increasing, which suggests that few of us truly took this pledge to heart.

When I think about the problems and challenges of global warming, I focus above all on how it will influence my comfort and lifestyle. Almost every day, the media report on disasters caused by anomalies that have occurred due to climate change. Global warming is happening, and we don't fully appreciate that this fact has an impact on our mental health. Psychologists all over the world are dealing with symptoms of climate depression and spiralling anxiety about the future. As time goes by, we can only expect this problem to grow.

During my gloomy daily searches, I've found a number of interesting studies of Inuit communities. In polar regions, climate change is happening significantly faster, and has already left a clear mark on the landscape. "We are people of the sea ice. If there's no more sea ice, how can we be people of the sea ice?" asked one member of a tribe.[148] The Inuit are deeply attached to their natural surroundings, so they perceive these transformations as a personal loss. The word *uggianaqtuq*, which used to mean the unpleasant feeling when someone close to you is behaving strangely or when you have a vague sense of missing something, has acquired a new meaning. Now *uggianaqtuq* refers to a situation where familiar things become unpredictable. The weather over the last few years has been very *uggianaqtuq*. Storms are long and violent, the ice is thin, and the amount of food available is in decline.[149] The sense of losing touch with a place causes depression; the sense of losing influence over reality weakens community ties. Frustration leads to increased consumption of drugs and alcohol.

In the West, the word *solastalgia* is gaining prominence. It refers to feelings of helplessness and distress caused by changes occurring in the environment. In a way it's related to the Inuit *uggianaqtuq*, since it hints at a sort of nostalgia, an ache for familiar but changing places. According to Eurobarometer statistics, 93 per cent of European Union citizens consider climate change a serious problem. In Poland specifically, the proportion of those worried by the situation is smaller but growing rapidly. 70 per cent of our society now consider the climate disaster to be a very serious problem, though we see terrorism, famine, poverty, lack of access to water and armed conflict as more significant issues facing the world.[150] I'm not sure how many of us grasp the fact that all these issues are connected.

What I find especially worrying is the idea that the climate catastrophe will strike in my lifetime, and certainly in my friends' children's lifetimes. I feel the need to act, but I don't know how, which leaves me feeling discouraged and apathetic. Sometimes I feel lonely and misunderstood, even among those closest to me. I struggle with the sense that my individual actions are insignificant and that no-one will hear my voice. I also know people who try to protect themselves by rejecting and denying the facts assailing them on all sides. What can I say? There are lots of different coping strategies for living with fear.

Therapists advise us to try to turn that fear into a motivating force. It's good to get in touch with people who share our anxieties. It brings relief. The feeling that you're being heard frees you from isolation. There are therapeutic benefits to getting involved, joining movements that aim to raise awareness of the consequences of climate change and exert pressure on governments. Participation and identification allow us to regain control over a situation that leaves us feeling helpless. And besides, running away from the latest alarming news won't help; keeping ourselves up to date allows us to adjust to what's happening.[151]

I've basically already adjusted to the sight of cranes in January. I watch as they circle in silence on a Sunday morning over Pole Mokotowskie park, a little too high for me to make out any details. They're flying east – is winter already over for them? I had to adapt quickly, because ten years ago those cranes would have sparked a small sensation, but nowadays it's a normal sight. Autumn runs straight into spring. Along the Vistula you can already see catkins on the pussy willows. Not yet fully unfurled or fluffy, still tucked away. A bird-watching friend writes that somewhere near Łowicz the spring flocks of geese are already gathering, about a month earlier than usual.

4.10 a.m. – obsessive thoughts, particularly when I can't sleep
7.00 p.m. – the sense of despair when I wake after two hours of sleep rather than a twenty-minute nap

When you start looking into the topic of zero waste you inevitably come across the name Bea Johnson. Her book, *Zero Waste Home*, left me with a growing sense of helplessness. Her proposed waste-free lifestyle felt not so much like good practice as a kind of religion or a game for the privileged. Every few pages, I'd put the book down with a scowl: being perfect is difficult, just as it's difficult to become a wealthy resident of California. Inventing reusable homemade alternatives to toilet paper is only feasible for someone who fundamentally has no other problems. People who hitchhike from their village to get to work and burn rubbish in winter because they can't afford eco-pea coal face bigger challenges than trying to bulk-buy nuts. And actually, their carbon footprint is probably lower than many of the angst-ridden residents of leafy Mokotów who fly to Asia in winter to follow the sun.

If you took the book as a guide, it might result in a breakdown. Producing no more than a kilogram of waste per year is simply impossible under current conditions. To live as Bea Johnson teaches, you'd have to give up your day job. I tetchily note down a few passages. For instance: "Once tied to expensive cosmetics, I now make do with the rich soap mentioned earlier to wash my body from head to toe." Or: "Long gone are the days when my bathroom drawers were filled with facial products. When I am in the mood for a facial, I simply head to the kitchen: everything I need is stored in my pantry."[152] It made me think of Marie-Antoinette dressed as a shepherdess tending her perfumed lambs and cows. This is advice for people who are bored with cosmetics, I thought.

And yet I kept going back to the book. As much as I derided the picture it painted of a saintly housekeeper, each time I picked it up again I'd jot something else down. I conscientiously took note of which house plants purify air, apparently according to analysis by NASA. We complain about smog all winter, but I don't want to buy an air purifier – I'm loath to acquire any more short-lived electronic devices. I bought a Sanseviera and a peace lily, and then I read that NASA's research isn't really valid for typical home conditions and it's better to simply unblock air vents. It doesn't matter, I'm happy to have those plants. I also copied out Bea's recipes for homemade detergents – and I'm sold: baking soda is surprisingly good for cleaning the bathroom. So what was my beef with Bea? It's not like she was forcing me to do anything.

8.00 a.m., 11.37 a.m., 4.12 p.m., 6.05 p.m., 8.50 p.m., 10.12 p.m. – procrastination and mindless scrolling through social media

For a long time, I took consolation from the thought that environmental neurosis takes a more aggressive form in others. I followed Facebook groups in which people indulged their fantasies about eco-friendly living, about a moral and ethical world. I too believe in conscious consumption and investing in things that last, but I couldn't escape the feeling that many members of the movement spend a great deal of time and energy chasing utopia, an ideal world in which they won't produce any rubbish, waste or carbon dioxide. That seemed naïve, but harmless enough; what annoyed me was the bragging matches between people trying to outdo one another in their dedication to the cause. It was like looking through a broken lens that erases both the details and the bigger picture.

Many people were intoxicated by their sense of moral superiority, berating those who can't live up to the orthodox ideal. Throwing out a pizza box or some stodgy pasta was enough to warrant a public lynching. I was irritated by their misleading insistence that these minor, innocuous acts of profligacy were so damaging to the planet. Do we really matter that much? I understand that we live in a narcissistic society, but let's be serious. People who take such satisfaction in lecturing others are putting on a spectacle for their own benefit. Flirting with their own reflection in the mirror. Shifting the blame for destroying the planet onto somebody else is probably a good way to make yourself feel better.

Does "eco-shaming" work? Perhaps in Sweden, where they have already coined neologisms to address environmental neurosis. *Flygskam* is the shame of travelling by plane and generating a large carbon footprint. To counter it, there is *Tagskryt*, pride in taking the train.[153] Once again, there's no denying the fact that rail travel is very often a pastime for the privileged few. International flights are frequently cheaper, to say nothing of how long the journey takes. Spending your time on a sustainable, mindful journey tends to be an option only for people who aren't in a hurry.

A while ago, I read a post that seemed funny at the time. A girl was asking where she could buy hydrogen peroxide in glass bottles, because she was trying to avoid plastic at all costs. I thought she was probably overestimating the impact of her consumer choices. Hydrogen peroxide isn't an essential, everyday product, after all. Selling it in properly recyclable HDPE bottles is not the most pressing environmental concern. And while you can probably order hydrogen peroxide in glass bottles over the internet, eco-friendly purchases often travel hundreds or thousands of kilometres from mail-order shops – which raises the question: what is it all for?

What was it that amused me about the hydrogen peroxide? I was laughing at myself. I mean, I'm afraid, too. Just like that girl, I feel a sense of impending doom and every day I wonder what I can do to stop the Earth from groaning under the weight of my footsteps. All these efforts remind me of a story from Mary Mapes Dodge's book *Hans Brinker, or the Silver Skates*, an American fantasy about life in a Dutch village. The main character, Hans, the son of a lock keeper, discovers one day that the dike is leaking. Without giving it too much thought, he plugs the hole with his finger and stays there all day and night, saving the country from disaster. It's said that only systemic changes can save the world, but how do you change the current economic model, the pressure for growth, enrichment, development, for upward trends on the graph? How do you put environmental interests first, ahead of economic concerns?

7.40 a.m., 5.20 p.m., 8.10 p.m. – visits to the corner shop

We have a cheeky nickname for our local shop: Dildo – which sort of echoes the original name. It isn't a big place, but it has everything its customers need: a small selection of vegetables and a large range of beers. I try to buy beer in returnable bottles. In Poland, you can only return bottles to the shop where you bought them, with proof of purchase, but as a regular customer at Dildo, I'm not fended off with the question, "Do you have the receipt for those?" This restrictive deposit system is obviously a problem. On the other hand, although unreturnable bottles can't be refilled, they can always be treated in a glass works; what matters is that they end up in the right bin.

Environmental neurosis really strikes when I see the shelves of packaging. I find it particularly disheartening when it looks like a manufacturer has insisted on a design that means it can only be used once and never recycled. As if the aim was to create useless rubbish. In recent years, there's been a proliferation of paper packaging with quirky plastic windows, making both materials practically impossible to recover, since separating them isn't economically viable. Plastic windows have now appeared on the packaging of my favourite pasta, probably to keep customers entertained by letting them spy on their spaghetti. I always make a point of separating the two parts when disposing of them.

Not far from the pasta is the flour. I find a packet that appears to be made of paper, but has a thin plastic lining on the inside. That must provide better protection against damp, but then why not make an honest, single-material packet out of polyethylene? A bag of flour looks the same as it did, but the difference is that it's now useless once empty. Any packaging that combines multiple materials needlessly complicates recycling and sometimes makes it impossible. We're fond of paper because it's natural and seems more environmentally friendly, but packaging manufacturers prey on that intuition, making multi-material packets covered in a coating that imitates the structure of paper. An unobservant consumer is unlikely to notice.

Even with unadulterated paper, things aren't that straightforward, incidentally. If it's going to be recycled, it has to be clean. And unfortunately, toxic substances contained in printing inks, such as MOAHs (mineral oil aromatic hydrocarbons) and MOSHs (mineral oil saturated hydrocarbons), spread easily on paper. Luckily, newspapers rarely come into contact with food nowadays. No-one uses them to wrap meat anymore, and slicing a tomato on a broadsheet is a disappearing element of the folklore of post-Soviet economy-class trains. And even polluted paper is useful for making corrugated cardboard.

A few steps away from the pasta and flour are the drinks bottles designed to appeal to children, with big caps and ugly, garish labels depicting the heroes of some popular film. The bottles themselves are usually made of PET, an easily recyclable material, but the optical sorting machines may not be able to correctly identify them and remove them from the conveyor belt in the sorting plant. The infrared beam will only see the heat-shrink wrap, often made of PVC (polyvinyl chloride), which forms a tight seal around the bottle and can't be treated. The same goes for plastic yoghurt pots. For the pot to stand a chance of being recycled, it's best to tear off the label.

Recyclers complain that multi-material packaging creates additional, unnecessary work separating it into homogeneous pieces. On my way to the counter, I pass yoghurts with cardboard sleeves, which get wet and clog the lines in recycling centres. There's a similar problem with packaging that includes a pump, like liquid soap. The metal springs that make the pistons pop back often jam treatment lines. The aluminium seals hidden in lids or caps are also problematic (especially as they aren't really necessary). Silicon seals and strong glues, if overlooked, can damage the quality of the recyclate).

8.00 p.m. – a casual dinner in town

Why do so many restaurants only serve foreign mineral water? It usually comes from Italy, for instance, "the official water of the Associazione Italiana Sommelier". It arrives here from Carisolo, sourced from Pra' dell'Era, and it tastes worse than the water from my tap. I can't stand it. Do we really have to regulate such idiocy with legislation? Isn't it objectively absurd for ordinary water to travel more than a thousand kilometres?

9.20 p.m. – before bed, a friend writes on Facebook that plastic packaging causes cancer

We'd like to have control over our lives, choosing only what is good and healthy for us. That's why we buy organic or eco-friendly food in good faith, but often all we're buying is adjectives and labels. We can rarely check the claims they make. What *is* real is the higher price. We also have our own ideas about what's environmentally friendly: dirty vegetables at the market seem somehow more authentic than shiny, neatly arranged ones in plastic packaging. But what do we know about how those market vegetables were grown? Do we know anything about the quantity and quality of pesticides used? About the earth from which they sprouted, or how they were transported?

I take some comfort in the knowledge that packaging and food contact materials have to be analysed in regulated laboratories before they can be unleashed on the general public. Food simulants, which imitate a given category of food, are placed inside the packaging being tested and then stored under various conditions. The labs issue a conformity declaration, confirming that the composition of the packaging is as declared by the manufacturer, and that the level of substance migration does not exceed certain limits. Still, it all sounds a little worrying, doesn't it?

Infringements of the rules are rare, says Adam Fotek from Hamilton, one of the large laboratories that study food safety. In recent years, there has been a recurring issue with a number of chemical substances leaching into food, including bisphenol A (BPA), a monomer used to make polycarbonate. The resulting plastics are both hard and shatterproof, which is why they are used to make objects that need to be resistant to shocks or to being dropped – babies' bottles, for example. The problem is that bisphenol disrupts the endocrine system and can cause infertility as well as breast and prostate cancers,[154] and it turned out that bisphenol migration into the food exceeded permitted levels.

The term "BPA-free" suddenly appeared on all sorts of reusable bottles – including some made of PET. Producing this plastic never involved bisphenol A, so mentioning its absence is perhaps overzealous, but it certainly calms down consumers. BPA is now used very little in packaging. It's still found in receipts, however, and is also a component in the inner lining of steel cans. Since 2011, there's been a ban on using the substance in polycarbonate items meant for children and babies. Since 2018, bisphenol A has also been prohibited in the layers of epoxy coating on products for that age group.

As part of the fashion for environmentally friendly materials, you can now buy many bamboo products made in China, such as bowls, plates and other kitchen items. And yet bamboo doesn't grow thick enough to allow for the production of a large chopping board from a single stem. A melamine formaldehyde resin is often used to glue pieces together. There have been cases of formaldehyde leakage that exceeds migration limits. In early 2019, a range of bamboo thermal mugs sold in one supermarket were scrapped for this reason. Formaldehyde is considered likely to be a carcinogen and harmful to the reproductive system. It also affects the respiratory tract and mucous membranes. But it's widely used as a preservative and a disinfectant, and in the production of synthetic resins, glues, cosmetics, chipboard and animal feed. Basically, it's everywhere.

8.10 p.m. – a third walk with the dog, more thoughts following that unpleasant afternoon awakening

For many years, there's been talk of the circular economy, which means rationally consuming resources, reducing the negative environmental impact of the products we produce, using materials and resources for as long as possible and minimising waste. There is very high support for introducing a circular economy: as many as 95.5 per cent[155] of those surveyed were in favour of replacing Poland's current model, a so-called linear economy, with a circular one. To me it looks like a distant mirage of a world in which common sense finally prevails.

Citizens will be educated. They will make sensible consumer choices. They will put the general interest above their own comfort. Industry and transport executives will pay attention to their carbon footprint and respect the environment.

At an industry meeting, an important figure in the world of the circular economy told a joke:

A rabbit fleeing a wolf bumps into an owl in the forest.

"Owl, you're so clever, tell me what to do – otherwise the wolf will eat me!" she says.

"Indeed . . ." says the owl, thinking hard. Then he says, "Rabbit, you need to turn into a tree."

"What a great idea!" says the rabbit, but she's immediately struck by doubt. "How do I do that?"

"Don't ask me! I'm just in charge of ideas," replies the owl.

4.00 p.m. – dusk
10.00 p.m. or more like 11.47 p.m. – sleep

The most beautiful time of day is the last half hour before dusk, when the winter sun gilds the breasts of passing rooks. The light creeps up the wall, then pauses for a moment on the old white poplar that looms over the neighbourhood. And then the sun disappears. It's somewhere over in Wola, as always. After a brief twilight, the street lamps come on. You can't see the stars in the sky, only the reflected light of the city, just as grey human faces shine in the glow of smartphone screens. Thousands of lamp posts, car headlights, billboards, backlit adverts, Christmas decorations, shop windows, illuminated office blocks and apartments, all contributing to light pollution, which is defined as when night skies are around 10 per cent brighter than under natural conditions.

This morning I saw another backlit advertising space being installed on a neighbouring building. Nobody asked me if I want to look at rock-bottom prices, new phones or insurance policies every day. But even without that new light source it was never dark on our street. And we're by no means alone in this – 83 per cent of the population experience light-polluted night skies. Natural darkness is only found in unpopulated areas, which in Europe means the northern reaches of Scandinavia and islands far from the mainland. More than one third of us can't see the Milky Way from where we live. And bear in mind that all these data are from six or seven years ago, when the most recent major analysis of satellite photos was carried out.[156] Light pollution is considered the fastest-growing form of pollution in the world. It's a problem that is not being given the attention it deserves, mostly because we don't realise the consequences of these changes.

Of course, artificial light disrupts ecosystems. It alters the biological rhythms of nocturnal organisms and modifies their habits. Light can confuse, attract or deter. For instance, small nocturnal mammals avoid well-lit areas because they're more likely to fall victim to predators there. When a source of artificial light appears in their habitat, they restrict their movements, spend less time searching for food and as a result their health suffers. In the case of migratory birds that navigate using the stars, many mistakenly pick brighter objects as their target. The first such case was noted back in 1886, when almost one thousand birds struck an illuminated tower in the town of Decatur in the United States. It's not unlike the story of the hundreds of baby turtles, freshly hatched on a Calabrian beach, who, instead of heading towards the sea, made a beeline for a nearby restaurant. The glow of its lights was more attractive than the moon's faint reflection on the waves.[157]

A few years ago, I saw an example of how light harms nature in the Żywiec Beskid mountains. It was late June, I was at a friend's wedding reception outside a newly built hotel, and right at the peak of the hill there was a floodlight shining in the darkness. At one point, the guests noticed dozens of glowing yellow dots circling the light. Fireflies from across the area had been drawn to it. The wedding guests watched, enchanted – many hadn't seen those insects for years. I was also enjoying the spectacle, not really grasping the tragedy that was unfolding before my eyes. Hundreds of insects, instead of dancing across meadows, instead of flying to a wedding, had hit upon a funeral. They were duped and fried by the floodlight.

Many species aren't bothered by artificial light and some actually welcome it. Diurnal spiders have taken to spinning their webs beneath street lamps.[158] Insects thronging around these same lights are also snatched up by fast-flying bats; the slower, less agile ones have to hunt for them in the dark where they're less exposed to predators. Such spots are getting harder to find. It's not just bats and spiders taking advantage. Peregrine falcons that have made their homes in cities also make use of urban lighting, having learned to hunt at night. This phenomenon was first recorded by cameras on the roof of Derby Cathedral. The local falcons were using the powerful city lights to prey on birds migrating at night. And in Warsaw, park blackbirds feeding near well-lit paths are a common sight.

Exposure to artificial light causes many disorders. Birds, for instance, begin their breeding and moulting seasons earlier. No wonder – artificial light hampers their ability to recognise seasons and the length of days. Under its influence, amphibians shorten their mating season to hasten reproduction. They are less discerning when it comes to selecting a partner, even though making the right choice is crucial for the continued transmission of their genes. Many animals subjected to artificial light develop disorders in melatonin production, the hormone responsible for regulating an organism's sleep–wake cycle. Disrupting this mechanism can lead to lowered resilience, reduced ability to fight infections, generally faster ageing of the body and increased incidence of cancers. Studies on humans exposed to artificial light at night (those who work late shifts) show a significantly increased risk of breast and prostate cancers.

It's dark outside but not pitch black, the light of my neighbour's television is coming through the window. These obsessive thoughts gradually fade from my mind and all I can hear is the quiet hum of the fridge, while the small diode on my bike-light charger shines next to my bed. The laptop on standby flickers from below. Water heated by energy from a coal power plant trickles inside the radiator. It's hard to sleep in the shadow of impending catastrophe. Every day brings new research, ever gloomier forecasts, shocking changes in perspective. Every day the gulf between what we always thought was the case and what is really happening gets wider. It's hard to keep up, and sometimes it's tempting to plug your ears and cover your eyes and mouth. Every day I fall asleep with the thought that perhaps tomorrow something will happen to give me hope. Perhaps when I wake up, the whole world will too.

MAY 3, 2020

5.18 p.m., 52°13'09.2" N, 21°02'25.8" E

It's early May, 2020. Single-use gloves and masks have become part of our daily lives. And part of our waste landscape. Today of all days, just before I submit this book for editing, I come across a photo by the ornithologist, Fatima Hayatli. Fatima took a picture in Warsaw of a nest with a disposable facemask woven among the twigs. A coot is sitting inside it, and four chicks peek out from under her. It's not just the mask that's out of place here. Normally, coots construct their mound-shaped nests from aquatic plants, then place bits of dry calamus or reeds on top and line the inside with soft bulrushes. But the only thing this one can gather from the concrete-lined canal that's cleared every year are branches from the linden trees growing by the bank. There's a plastic canopy in place of the reed shelter and a wooden board instead of a mound. So perhaps the facemask was the next best thing to bulrushes. That's how coots live in the Anthropocene.

*

The virus has paralysed daily life, from India, to Alabama, to the village of Tomczyce. Plans for spring birdwatching walks, trips and author talks have vanished from my calendar as if written in invisible ink. This book was supposed to be published in late February or early March, then it was delayed until April. It skipped lithely over May and will now come out in June. Like the Chief Public Health Inspector Jarosław Pinkas, who at the end of January assured us that "the virus is currently in China – that's quite a long way from here", I didn't think the problems of a certain Chinese city would become the problems of the entire world within just a few months. Clearly, the human intellect still hasn't come to terms with the consequences of globalisation.

In my own, fairly comfortable case, every day has become a weird Sunday afternoon, with all its soporific hallmarks. My street, with its endless traffic jams and the ever-present stink of fumes, has become rather sleepy. Not that I'm in any danger of dozing off, because every five minutes an ambulance rushes past, its siren blaring. Social life has moved online, where conspiracy theories are spreading like viruses: theories about secret biological warfare, how this all serves such-and-such's interests, that it's 5G, the World Government. I look on helplessly as my friends share this hogwash. The internet really is a huge, unsegregated, toxic wastebin.

People say nature is breathing a sigh of relief, that turtles have returned to Indian beaches and boars are back in Tarchomin. Who would get worked up about waste at a time like this? And yet, the recycling market is facing a severe crisis. 90 per cent of French sites[159] have shut because of the pandemic. In Germany consumers have stopped returning PET bottles to deposit machines.[160] In the United Kingdom, there are shortages of cardboard because collection points for wastepaper have closed.[161] At this inauspicious time for major expenditure, big businesses in the US plastics industry have asked for one billion dollars of support for investment in recycling infrastructure.[162] Levels of separated refuse collection are falling; many countries have relaxed rules on waste segregation. On top of this, the ongoing trade war between Saudi Arabia and Russia means the price of oil is dropping. Both countries increased supply, but demand fell at the same time – air traffic has almost ceased, logistics have slowed and ordinary road use has plummeted dramatically. The price per barrel is at a record low. Inexpensive oil means inexpensive raw materials, which in turn means no incentive to buy recovered materials. Producing new plastic is cheap.

*

In Poland, you have to do your shopping in single-use gloves. Every day I find a few different new specimens outside my building. Transparent polyethylene ones. Nitrile, latex, vinyl and God-knows-what else. White, green and blue. On windy days I see the transparent plastic ones flying high up above the rooftops, as I look out for migrating birds. I've already seen a stork, then a crane, a squadron of cormorants, some buzzards, but I'm much more likely to see these supermarket gloves. Once they've been casually dropped on the pavement, one gust of wind can sweep them up and carry them for miles.

Will all the rubbish settle into a clear geological stratum

from this period? As early as March 12, I was reading alarming news from Hong Kong. Single-use masks from that administrative region of seven million inhabitants were washing up on nearby beaches.[163] The problem will grow, because almost three quarters of waste there is sent to landfill. In all likelihood, a lot of it will be washed away with surface water into the ocean. Masks float well. They can easily travel the thirty kilometres separating Hong Kong from the uninhabited Soko islands.

Pandemic waste in seas and oceans is a problem for the entire world. At the end of March, the archaeologist Stein Farstadvoll, who lives in Tromsø, Norway, found a bottle of disinfectant washed up on the beach. The writing on the label was in Norwegian, the bottle probably hadn't been adrift for long. A month later, Stein found a sea-beaten single-use mask manufactured by the Norwegian company Granberg and some gloves. In the next few years, this kind of litter will travel hundreds or perhaps thousands of kilometres and be found on coasts around the world.

As early as mid-March, I found pictures online of Poland's first pandemic litter. A discarded surgical mask was basking in a meadow near the Chojnik mountain in the Karkonosze National Park. I didn't expect such sights to become commonplace and normal within a month. Who dropped their mask there, off the beaten track? Were they already infected or afraid of infection? (It wasn't mandatory to wear masks outside at that time.) Unfortunately, concern for your own health rarely goes hand in hand with the health of others. Abandoned protective equipment can spread infection. Somebody has to clear up the discarded masks, gloves and tissues, after all.

*

Research[164,165] confirms that the virus's resilience depends on the surface on which it is found. SARS-CoV-2 vanishes from cloth and treated wood. It can survive up to three hours on surfaces like printed paper, a few days on smooth surfaces such as plastic or steel, and up to a week on the inner lining of a mask. At the same time, the virus is easily killed by standard disinfectants. But let's remember that these studies are carried out in controlled laboratory conditions, their results can't automatically be translated into the everyday landscape of shops, trams or church pews. And finally – traces of the virus don't necessarily mean there are pathogens capable of reproduction (that is, of spreading).

All over the world, the waste industry is wary of infections among its employees. It's suspected that municipal waste can transmit the virus. Workers being required to quarantine could delay or prevent waste collection; in many cases it could even lead to the closure of entire facilities. The European Commission[166] states that there is no proof of waste playing a significant role in virus transmission. It does, however, recommend taking precautions: personal protective equipment (gloves and masks), disinfection and social distancing. In the United Kingdom, sick people and those in quarantine are asked to double-bag their waste and keep it outside their homes for three days. Garden and bulky waste is not collected. Refuse from self-isolating households is incinerated.

Typically for Poland, mild confusion reigns. In March, waste management companies called on the prime minister to issue guidelines for the duration of the pandemic. There were no procedures, for instance, for how to collect waste from residents infected with coronavirus. After a few weeks, the climate minister and Chief Public Health Inspector obliged. The document was vague and non-binding. "It is recommended that separated waste be stored, where possible, for nine days before being sent for treatment." "It is recommended" means no-one will mind if it doesn't happen. The obligation on waste management companies was to "provide bags in a specific colour and/or marked with a symbol (such as the letter 'C')", but only "to the extent possible".

In practice, everyone did the best they could. In Zawiercie, red bins labelled "COVID-19" were placed on streets.[167] It didn't occur to anyone that this stigmatised the sick; certain residents quickly identified the apartment in which an infected person (a doctor) was self-isolating. Or there was the proposal from Bogdan Zieliński, prefect of Wysokie Mazowieckie county, to tag the homes of those infected with coronavirus.[168] He justified this medieval idea on the grounds of public health. Fortunately, nobody took it seriously, though stories of attacks and harassment of healthcare workers have come to light from across Poland. When humans are afraid, they become a little less humane.

I also have first-hand anecdotal evidence. I know a doctor who had coronavirus. Since she had good reason to expect the test to be positive (she had treated an infected patient), she didn't wait for the outcome but self-isolated voluntarily. She received the test result after eight days. From that point, she had to undergo a mandatory two-week quarantine. Although she lived alone, the police were only really concerned with checking whether she was at home and not dead – sometimes several times a day. Nobody offered to bring food or take out the bins. After two weeks of isolation, the National Public Health Inspectorate was supposed to test her to confirm she was no longer infected. Inspectorate workers dressed like less entertaining versions of the Teletubbies went awkwardly from door to door, floor by floor, asking if Ms X lived there. These alien visitors understandably stoked fear among her neighbours. After two negative tests and almost a month locked inside her flat, she was finally released from quarantine and could dispose of the rubbish she'd collected on her balcony.

*

During the pandemic we've produced less waste overall and its composition has changed. Fearing uncertainty and lean months ahead, people have avoided making unnecessary purchases. After the initial panic-induced plundering of toilet paper, rice and yeast, consumers have focused on more essential items. And to a large extent, shopping has moved online. As always, the excessive packaging riles me. Small objects wrapped up as if they were glass pianos. Sales of razors, sewing machines and dumb bells for home workouts have increased. Naturally, we're not spending money on travel; Booking.com, railway companies and Uber have seen huge drops in revenue.

Disposables are the real winners. From early March, you could no longer buy coffee in your own mug at Starbucks. The same was true on board Great Western Railway trains and in most takeaways (perhaps all of them?). The measure was justified by health and safety concerns. Lobbyists sprang into action; packaging companies claimed plastic containers were safer. Producers of single-use cutlery and plates pricked up their ears. A letter[169] from the European Plastics Converters landed on the European Commission's desk, calling for the ban on single-use items to be delayed. But, of course, single-use doesn't mean sterile.

The American state of New Hampshire took its precautions to an extreme, banning the use of reusable bags for fear that they might transmit the virus. The plastics sector had already smelled blood. In a letter[170] to the Department of Health, the Plastics Industry Association called "bans on these [single-use] products [. . .] a public safety risk". To argue their case, they referred to research from 2011 that confirmed the presence of bacteria on reusable bags – research[171] commissioned by the American Chemistry Council (an association of chemical and fuel companies). However, the studies I mentioned above show the virus survives for longer on plastic bags than on cotton ones.

We're told that the world is changing before our eyes and that the pandemic may have a lasting impact on our daily lives, on our lifestyles, and certainly on our ability to travel. We're already replacing smiles, hidden behind a mask, with a blink of the eyes, and we avoid each other as if repelled by an invisible aura, like magnets with the same poles. There's the risk that viruses become a convenient smokescreen, an excuse to postpone the introduction of essential changes, like moving away from fossil fuels. The pandemic may be a pretext for turning the screw and increasing social control. The truth is, we simply don't know.

ACKNOWLEDGEMENTS

This book contains an extended, updated version of the column "Nineteen Cartridge Cases" about hunting ammunition, which appeared in *Dwutygodnik*. Many of the ideas, words, and sometimes sentences about Dichlorvos, plastic bags, throwaway products, balloons and paper lanterns, marine waste patches, straws and so on were first found in my columns published in the "Litter of the Issue" series. I thank editor-in-chief Zofia Król for her patience and encouragement. Without *Dwutygodnik*, *What We Leave Behind* would not exist.

These people have helped me at different stages of this book:
Michał Łukawski
Marta Krawczyk
Marta Sapała
and also
Piotr Barczak, Mirosław Baściuk, Piotr Bednarek, Dr. Wojciech Brzeziński, Jurek Cependa, Dr. Agnieszka Dąbrowska, Dr. Wojciech Fabianowski, Adam Fotek, Przemysław Gumułka, Fatima Hayatli, Dr. Krzysztof Kolenda, Edyta Konik, Dominik

Krupiński, Diana Lelonek, Jagna Lewandowska, Roman Miecznikowski, Aleksandra Niewczas, Dr. Grzegorz Pac, Dr. Michał Paczkowski, Dr. Włodzimierz Pessel, Dr. Igor Piotrowski, Matthieu Rama, Przemysław Stolarz, Prof. Piotr Tryjanowski, Robert Wawrzonek, Marta Wiśniewska, Karol Wójcik and Szymon Rytel, Zosia and Łukasz, Paweł Ponury Żukowski
    and
Paweł Goźliński.

My thanks to the companies MPO, the Municipal Waste Management Enterprise, and to BYŚ Wojciech Byśkiniewicz for the opportunity to observe their work.

I'd also like to thank my companions in life:
    D and M (+ F and M), my parents, Wiktor, Max, Małgo and Boru, Marta and Szymon, Jot and Miły & Misia, Maciek.

NOTES

1   S. Gibbens, "Plastic proliferates at the bottom of the world's deepest ocean trench", 14/05/2019, https://www.nationalgeographic.com/science/article/plastic-bag-mariana-trench-pollution-science-spd, [accessed: 14/04/2021].

2   United Nations Environment Programme, *Single-Use Plastic: A Roadmap for Sustainability*, 2018 https://wedocs.unep.org/handle/20.500.11822/25496, [accessed: 15/04/2021].

3   A. Vaughan, "Morrisons' paper bag switch is bad for global warming, say critics", *Guardian*, 25/06/2018, https://www.theguardian.com/business/2018/jun/25/morrisons-paper-bag-switch-is-bad-for-global-warming-say-environment-agency, [accessed: 15/04/2021].

4   "Tesco says it WON'T be joining other supermarkets phasing out plastic bags for loose fruit and veg because papers ones have a LARGER carbon footprint", *Daily Mail,* 03/09/2019 https://www.technologybreakingnews.com/2019/09/tesco-says-it-won-t-be-joining-other-supermarkets-phasing-out-plastic-bags-for-loose-fruit-and-veg-because-paper ones have-a-lar/, [accessed: 27/09/2021].

5   C. Mitrus, A. Zbyryt, "Wypływ polowań na ptaki i sposoby ograniczania ich negatywnego oddziaływania" [The influence of hunting on birds and how to reduce its negative impact (in Polish)], *Ornis Polonika*, 2015, no 56, pp.309–327.

6   *Przyrodniczo-ekonomiczna waloryzacja stawów rybnych w Polsce* [An Environmental Economic Evaluation of Fish Ponds in Poland (in Polish)], ed. K. A. Dobrowolski, Warsasw, 1995, p. 56.

7   Regulation of the Minister for Environment of 28 December 2009 on Hunting Permits (in Polish), http://prawo.sejm.gov.pl/isap.nsf/download.xsp/WDU20100030019/O/D20100019.pdf, [accessed: 27/09/2021].

8   P. Wylęgała, Ł. Ławicki, "Głowienka, czernica, cyraneczka, łyska – stan populacji w Polsce i wpływ gospodarki łowieckiej. Opinia na potrzeby Polskiego Komitetu Krajowego IUCN" [The pochard, tufted duck, teal, coot – the state of the population in Poland and influence of game hunting. An opinion for the Polish National Committee of the IUCN (in Polish)], Poznań, 2019.

9   "Obwody łowieckie. Analiza gatunku" [Hunting Areas: a species analysis (in Polish)] https://niechzyja.pl/baza_wiedzy/dane-statystyczne-i-analizy/obwody-lowieckie-analiza-gatunku/, [accessed: 20/04/2021].

10   D. Wiehle, "Śmiertelność ptaków w wyniku polowań na Stawach Zatorskich w obszarze Natura 2000 ‹Dolina Dolnej Skawy›" [Bird mortality from hunting at the Zatorskie Ponds in the 'Lower Skawa Valley' Natura 2000 area (in Polish)], *Chrońmy Przyrodę Ojczystą*, 2016, no 72(2), p. 110.

11   Ł. Bińkowski, "Skażenie ptaków wodnych ołowiem w Polsce" [Lead contamination in waterfowl in Poland (in Polish)], *Brać Łowieka*, 2011, no 8.

12   R. Mateo, R. Toledo, "Lead Poisoning in Wild Birds in Europe and the Regulations adopted by different countries", *Ingestion of Lead from Spent Ammunition: Implications for Wildlife and Humans*, 01/01/2009, https://researchgate.net/publication/238734173, [accessed: 20/04/2021].

13   M. A. Tranel, R. O. Kimmel, "Impacts of lead ammunition on wildlife, the environment, and human health – a literature review and implications for Minnesota", *Ingestion of Lead from Spent Ammunition: Implications for Wildlife and Humans*, 2009 https://www.peregrinefund.org/subsites/conference-lead/PDF/0307%20Tranel.pdf, [accessed: 20/04/2021].

14   California Department of Fish and Wildlife, "Nonlead Ammunition in California", https://wildlife.ca.gov/hunting/nonlead-ammunition, [accessed: 20/04/2021].

15   Binkowski et al., "Histopathology of liver and kidneys of wild living Mallards *Anas platyrhynchos* and Coots *Fulica atra* with considerable concentrations of lead and cadmium", *Science of the Total Environment*, 2013, vol. 450–451, pp. 326–333, https://www.sciencedirect.com/science/article/abs/pii/S0048969713001654, [accessed: 23/04/2021].

16   R. Mateo, op. cit.

17   Stacja Badawcza PZŁ Czeplin, "Zestawienia danych sprawozdawczości łowieckiej 2019 rok" [Overview of hunting reporting data 2019 (in Polish)], 2019, http://www.czempin.pzlow.pl/palio/html.wmedia?_Instance=pzl_www&_Connector=palio&_ID=5849&_CheckSum=1038513142, [accessed: 23/04/2021].

18   "Wezwanie" [Summons (in Polish)], 20/08/2018, https://www.pzlow.pl/attachments/article/463/wezwanie.pdf, [accessed: 23/04/2021].

19   BirdLife International, *The Killing*, 08/2015, https://www.birdlife.org/sites/default/files/attachments/01-28_low.pdf, [accessed: 23/04/2021].

20   Committee Against Bird Slaughter, "Firing numbers in bird hunting in Europe", 30/10/2018, https://laczanasptaki.pl/wp-content/uploads/2018/11/CABS-Hunting-in-Europe-2017.pdf, [accessed: 23/04/2021].

21   Directive 2009/147/EC of the European Parliament and of the Council of 30 November 2009 on the conservation of wild birds https://eur-lex.europa.eu/legal-content/EN/TXT/?uri=CELEX%3A32009L0147, [accessed: 23/04/2021].

22   J. M. Arnemo et al., "Health and Environmental Risks from Lead-based Ammunition: Science Versus Socio-Politics", *Ecohealth* 13(4), pp. 618–622, 2016, https://doi.org/10.1007/s10393-016-1177-x

23   Group of Scientists, "Wildlife and Human Health Risks from Lead-based Ammunition in Europe. A Consensus Statement by Scientists", 2014,

http://www.zoo.cam.ac.uk/leadammunitionstatement/, [accessed: 23/04/2021].

24   M. A. Riva, A. Lafranconi, M. I. D'Orso, G. Cesana, "Lead Poisoning: Historical Aspects of a Paradigmatic 'Occupational and Environmental Disease'", *Saf Health Work* 2012, no 3(1), pp. 11–16, https://www.ncbi.nlm.nih.gov/pmc/articles/PMC3430923/, [accessed: 23/04/2021].

25   M. O. Aneni, "Lead Poisoning in Ancient Rome" (2018) https://www.research-gate.net/publication/325023100_Lead_Poisoning_in_Ancient_Rome, [accessed: 23/04/2021].

26   M. A. Tranel, R. O. Kimmel, op. cit.

27   World Health Organization, "Lead Poisoning and Health", 23/08/2019, https://www.who.int/en/news-room/fact-sheets/detail/lead-poisoning-and-health, [accessed: 23/04/2021].

28   *CIC Workshop Report Sustainable Hunting Ammunition*, ed. N Kanstrup, Aarhus 2009, http://www.cic-wildlife.org/wp-content/uploads/2013/04/CIC_Sustainable_Hunting_Ammunition_Workshop_Report_low_res.pdf, [accessed: 23/04/2021].

29   European Waste Framework Directive, Directive 2008/98/EC.

30   *Historyczne kopalnie – dzieło przyrody, sztuka człowieka* [Historic mines – works of nature, art of man (in Polish)], ed. B. Furmanik, A. Wawrzyńczuk, K. Piotrowska, Warsaw 2016.

31   D. Adamczyk, *Silber und Macht: Fernhandel, Tribute und die piastische Herrschafts-bildung in nordosteuropäischer Perspektive (800–1100)* [Silver and Power: tributes and long-distance trade in the formation of the Polish state and neighbouring countries, 800–1100 (in German)], Deutsches Historisches Institut Warschau 28.) Wiesbaden: Harrassowitz, 2014.

32   S. Strasser, *Waste and Want: A Social History of Trash*, New York 2000, pp. 71–72.

33   Ibidem, pp. 70–72.

34   Ibidem, p. 26.

35   Ibidem, pp. 170–171.

36   K. Ashenberg, *The Dirt on Clean: An Unsanitized History*, Canada, 2007, p. 128.

37   R. Marchand, *Advertising the American Dream: Making Way for Modernity, 1920–1940*, Berkeley, Los Angeles, London, 1985, p. 157.

38   C. Frederick, *The New Housekeeping: Efficiency Studies in Home Management*, Garden City, N.Y., 1913, p. 51.

39   G. Esperdy, *Modernizing Main Street Architecture and Consumer Culture in the New Deal*, Chicago, 2008, p. 156.

40   S. Strasser, op. cit. pp. 209–211.

41   Ibidem, pp. 233–259.

42   *Modern Plastics, Catalog Directory*, October 1936, p. 110, in C. and P. Fiell, *Plastic Dreams: Synthetic Visions in Design*, 2009.

43   R. Barthes, "Plastic" in *Mythologies*, translated by Annette Lavers, New York, 1972, p. 99.

44   B. Cosgrove, "Throwaway Living: When Tossing Out Everything Was All the Rage", *Time*, 15/05/2014.

45   J. J. Kolstad, E. T. H. Vink, B. De Wilde, L. Debeer, "Assessment of anaerobic degradation of Ingeo polylactides under accelerated land-fill conditions", *Polymer Degradation and Stability*, 2012, vol. 97, no 7, pp. 1131–1141, https://www.sciencedirect.com/science/article/pii/ S0141391012001413?via%3Dihub, [accessed: 06/05/2021].

46   M. Zafeiridou, N. S. Hopkinson, N. Voulvoulis, "Cigarette Smoking: An Assessment of Tobacco's Global Environmental Footprint Across its Entire Supply Chain", *Environmental Science and Technology*, 2018, 52 (15), pp. 8087–8094, https://pubs.acs.org/doi/10.1021/acs.est.8b01533, [accessed: 06/05/2021].

47   Ocean Conservancy, *The Beach and Beyond: 2019 Report*, https://oceanconservancy.org/wp-content/uploads/2019/09/Final-2019- ICC-Report.pdf, [accessed: 06/05/2021].

48   University Health Services, UC Berkeley, "Facts about Cigarette Butts and Smoke", https://uhs.berkeley.edu/tobaccofacts, [accessed: 06/05/2021].

49   M. Kaplan, "City birds use cigarette butts to smoke out parasites", *Nature* (2012), https://doi.org/10.1038/nature.2012.11952.

50   M. Suárez-Rodríguez, C. Macías Garcia, "There is no such a thing as a free cigarette; lining nests with discarded butts brings short-term benefits, but causes toxic damage", *Journal of Evolutionary Biology*, 2014, https:// onlinelibrary.wiley.com/doi/full/10.1111/jeb.12531, [accessed: 06/05/2021].

51   L. Lebreton, B. Slat, F. Ferrari, "Evidence that the Great Pacific Garbage Patch is rapidly accumulating plastic", *Scientific Reports*, 2018, no 8, https://doi.org/10.1038/s41598-018-22939-w

52   C. Schmidt, T. Krauth, S. Wagner, "Export of Plastic Debris by Rivers into the Sea", *Environment Science & Technology*, 2017, vol. 51, issue 21, https://doi.org/10.1021/acs.est.7b023868

53   J. Kröll, "Millionen Plastikteilchen der 'MSC Zoe' angespült – wie konnte das nur passieren?" [Millions of plastic particles washed up from the 'MSC Zoe' – how could it happen?] *Stern*, 09/03/2019, https://www.stern.de/panorama/ weltgeschehen/millionen-plastikteilchen-an-straenden--wie-war-der--msc- zoe--unfall-nur-moeglich-8614088.html, [accessed: 11/05/2021].

54   D. Hohn, "The great escape: the bath toys that swam the Pacific", *Guardian*, 12/02/2012, https://www.theguardian.com/environment/2012/feb/12/ great-escape-bath-toys-pacific, [accessed: 11/05/2021].

55   "The Big Interview: Professor Richard Thompson", *Invenite*, issue 2, University of Plymouth, https://www.plymouth.ac.uk/alumni-friends/invenite/issue-2/ the-big-interview-professor-richard-thompson, [accessed: 18/05/2021].

56   T. Gouin et al., "A thermodynamic approach for assessing the environmental exposure of chemicals absorbed to microplastic", *Science and Technology*, 2011, vol. 45, issue 4, https://pubs.acs.org/doi/10.1021/es1032025, [accessed: 18/05/2021].

57   A. A. Koelmans, E. Besseling, W. J. Shim, "Nanoplastics in the Aquatic Environment: Critical Review", in *Marine Anthropogenic Litter*, ed. M. Bergmann, L. Gutow, M. Klages, Cham, 2015, https://doi.org/10.1007/978-3-319-16510-3.

58   S. C. Votier, K. Archibald, G. Morgan, L. Morgan, "The use of plastic debris as nesting material by a colonial seabird and associated entanglement mortality", *Marine Pollution Bulletin* 2011, vol. 62, pp. 168–172, https://doi.org/10.1016/j.marpolbul.2010.11.009

59   S. Osborne, "Dead whale washes up on Indonesian beach with over 1,000 pieces of plastic in its stomach", *Independent*, 20/11/2018, https://www.independent.co.uk/news/world/asia/dead-whale-plastic-stomach-indonesia-ocean-beach-kapota-usland-wakatobi-national-park-a8642731.html, [accessed: 18/05/2021].

60   "Plastic Particles in Fulmar Stomachs in the North Sea", OSPAR, 2017, https://oap.ospar.org/en/ospar-assessments/intermediate-assessment-2017/pressures-human-activities/marine-litter/plastic-particles-fulmar-stomachs-north-sea/, [accessed: 18/05/2021].

61   World Economic Forum, "The New Plastics Economy: Rethinking the Future of Plastics", http://www3.weforum.org/docs/WEF_The_New_Plastics_Economy.pdf, [accessed: 19/05/2021].

62   KIMO, "Fishing for Litter Scotland: Final Report 2014-2017", https://www.fishingforlitter.org.uk/assets/file/FFLS%202014%20-17%20Final%20Report.pdf, [accessed: 19/05/2021].

63   5 Gyres Institute, "Why the Ocean Clean Up Project Won't Save Our Seas", 09/09/2015, http://www.planetexperts.com/why-the-ocean-clean-up-project-wont-save-our-seas/, [accessed: 19/05/2021].

64   Heinrich Böll Stiftung, "Plastic Atlas: Facts and Figures about the world of synthetic polymers", 2019, https://www.boell.de/sites/default/files/2019-11/Plastic%20Atlas%202019.pdf, [accessed: 19/05/2021].

65   Z. Herbert, *Still Life with a Bridle* (translated by John and Bogdana Carpenter, New York, 1991.

66   PlasticsEurope, "Plastics – the Facts 2019. An analysis of European plastics production, demand and waste data", 2019, https://www.plasticseurope.org/en/resources/publications/1804-plastics-facts-2019, [accessed: 21/05/2021].

67   "Larwy, które przetwarzają odpady, np. Styropian" [Larvae that digest waste such as Styrofoam (in Polish)], 24/03/2017, https://kopalniawiedzy.pl/macznik-mlynarek-chrzaszcz-larwa-styropian-polistyren-odpady-Magdalena-Bozek-Radoslaw-Rutkowski,26188, [accessed: 21/05/2021].

68   Скандал в ОАО "Алтайхимпром", https://www.amic.ru/photo/1269/, [accessed: 24/05/2021].

69   M. F. Bouchard, D. C. Bellinger, R. O. Wright, M. G. Weisskopf, "Attention Deficit/Hyperactivity Disorder and Urinary Metabolics of Organophosphate Pesticides", *Pediatrics* 2010, vol. 125, no 6, https://pediatrics.aappublications.

org/content/pediatrics/125/6/e1270.full-text.pdf, [accessed: 24/05/2021].

70   AW, "Ponad 320 kg śmieci na jednego mieszkańca. Co się z nimi dzieje? Zobacz raport GUS" [Over 320kg of rubbish per resident. What happens to it? Read the Office for Statistics' report (in Polish)], *Portal Samorządowy*, 19/11/2019, https://www.portalsamorzadowy.pl/gospodarka-komunalna/ponad-320-kg-smieci-na-jednego-mieszkanca-co-sie-z-nimi-dzieje-zobacz-raport-gus,135161.html, [accessed: 27/05/2021].

71   Eurostat, "Municipal Waste Statistics", https://ec.europa.eu/eurostat/statistics-explained/index.php/Municipal_waste_statistics#Municipal_waste_generation, [accessed: 27/05/2021].

72   J. Pinkser, "Americans Are Weirdly Obsessed with Paper Towels", *Atlantic*, 10/12/2018, https://www.theatlantic.com/family/archive/2018/12/paper-towels-us-use-consume/577672/, [accessed: 01/06/2021].

73   Article 4, Directive 2008/98/EC of the European Parliament and of the Council of 19 November 2008 on waste and repealing certain directives, OJ L 312 22.11.2008, p. 3, http://data.europa.eu/eli/dir/2008/98/2018-07-05, [accessed: 01/06/2021].

74   Conversio Market & Strategy GmbH, "Global Plastics Flow 2018", https://www.carboliq.com/pdf/19_conversio_global_plastics_flow_2018_summary.pdf, [accessed: 01/06/2021].

75   D. Jones, "Criminal 'trash mafia' gangs burn Tesco bags and M&S packaging that were sent for recycling in Britain 1,600 miles away on a rubbish tip in Poland", *Daily Mail*, 03/08/2018, https://www.dailymail.co.uk/news/article-6024951/British-recycled-waste-dumped-toxic-Polish-tip.html, [accessed: 01/06/2021].

76   BBC Reality check team, "Why some countries are shipping back plastic waste", https://www.bbc.co.uk/news/world-48444874, 02/06/2019, [accessed: 01/06/2021].

77   J. Dell, "157,000 Shipping Containers of U.S. Plastic Waste Exported to Countries with Poor Waste Management in 2018", 06/03/2019, https://www.plasticpollutioncoalition.org/blog/2019/3/6/157000-shipping-containers-of-us-plastic-waste-exported-to-countries-with-poor-waste-management-in-2018, [accessed: 01/06/2021].

78   Kittel, B, *System: Jak mafia zarabia na śmieciach* [The System: How the Mafia Profits from Waste (in Polish)], Warsaw, 2013.

79   "Jak Polska poradzi sobie z ambitnymi celami w zakresie recyklingu?" [How will Poland deal with ambitious recycling targets? (in Polish)], 08/10/2019, https://www.money.pl/gospodarka/jak-polska-poradzi-sobie-z-ambitnymi-celami-w-zakresie-recyklingu-6432863317022337a.html [accessed: 27/09/2021].

80   V. Makarenko, *Tajne służby kapitalizmu* [Capitalism's Secret Services (in Polish)], Kraków, 2008.

81   J. Reilly, *The Ballad of A.J. Weberman*, video recording, 1969, https://mediaburn.org/video/ballad-of-a-j-weberman-2/, [accessed: 02/06/2021].

82   T. Szaky, *Outsmart Waste: The Modern Idea of Garbage and How to Think Our Way Out of It*, San Francisco, 2014.

83   M. Krawczyk, "Popieram recyckling – segreguję odpady" [I support recycling – I separate my waste (in Polish)], a leaflet, Warsaw, p. 21.

84   R. Geyer, J. Jambeck, K. L. Law, "Production, use and fate of all plastics ever made", *Science Advances 3(7)*, July 2017, https://www.researchgate.net/publication/318567844_Production_use_and_fate_of_all_plastics_ever_made, [accessed: 07/06/2021].

85   C. Nicastro, "Sweden's recycling (d)evolution", https://zerowasteeurope.eu/2017/06/swedens-recycling-devolution/, 20/06/2017, [accessed: 07/06/2021].

86   European Commission, *The Environmental Implementation Review 2019. Country Report Denmark,* https://ec.europa.eu/environment/eir/pdf/report_dk_en.pdf, [accessed: 07/06/2021].

87   Danish Environment Ministry, "I/S Norfors, Affaldsforbrænding: Meddelelse af vilkår ved påbud" [I/S Norfors Waste incineration: notification of conditions to be met (in Danish)], 03/07/2019, https://mst.dk/service/annoncering/annoncearkiv/2019/jul/is-norfors-affaldsforbraending/, [accessed: 07/06/2021].

88   A. Arkenbout, *Hidden Emissions: A Story from the Netherlands. Case Study*, Zero Waste Europe, 2018, https://zerowasteeurope.eu/wp-content/uploads/2018/11/NetherlandsCS-FNL.pdf, [accessed: 07/06/2021].

89   Raad van State, "Uitspraak", 29/05/2019 https://uitspraken.rechtspraak.nl/inziendocument?id=ECLI:NL:RVS:2019:1737, [accessed: 15/08/2022].

90   L. Roman, B. D. Hardesty, M. A. Hindell et al., "A quantitative analysis linking seabird mortality and marine debris ingestion", *Scientific Reports*, 2019, no 9, https://doi.org/10.1038/s41598-018-36585-9.

91   Shs/aw, "German police investigate three suspects after Krefeld zoo fire", *DW*, 02/01/2020, https://www.dw.com/en/german-police-investigate-three-suspects-after-krefeld-zoo-fire/a-51859899, [accessed: 07/06/2021].

92   D. T. Gwynne, D. C. F. Rentz, "Beetles on the Bottle: Male Buprestids Mistake Stubbies for Females (Coleoptera)", *Journal of the Australian Entomological Society*, 1983, vol. 22. https://doi.org/10.1111/j.1440-6055.1983.tb01846.x

93   C. A. B. Mazine et al., "Disposable Containers as Larval Habitats for *Aedes aegypti* in a City with Regular Refuse Collection: A Study in Marilia, São Paulo State, Brazil", *ActaTropica*, 1996, no 62, pp. 1–13.

94   M. Michlewicz, P. Tryanowski, "Anthropogenic Waste Products as Preferred Nest Sites for *Myrmica rubra* (L.) (Hymenoptera, Formicidae)", *Journal of Hymenoptera Research*, 2017, no 57, pp. 103–114. https://doi.org/10.3897/jhr.57.12491

95   Z. A. Jagiello et al., "Factors determining the occurrence of anthropogenic materials in nests of the white stork *Ciconia ciconia*", *Environmental Science and Pollution Research*, 2018, no 25, pp. 14726–14733, https://doi.org/10.1007/s11356-018-1626-x

96   F. Sergio et al., "Raptor Nest Decorations are a Reliable Threat Against Conspecifics", *Science*, 2011, no 331, https://doi.org/10.1126/science.1199422

97   L. Kisup et al., "Plastic Marine Debris used as Nesting Materials of the Endangered Species Black-Faced Spoonbill Platalea minor Decreases by Conservation Activities", *Journal of the Korean Society for Marine Environment and Energy*, 2015, no 18, pp. 45–49, https://doi.org/10.7846/JKOSMEE. 2015.18.1.45

98   M. Antczak et al., "A New Material for Old Solutions – the Case of Plastic String Used in Great Grey Shrike nests", *Acta Ethologica*, 2010, no 13. https://doi.org/10.1007/s10211-010-0077-2

99   Ibidem.

100   Kulesza et al., "Nabici w butelkę" [Squeezed into a Bottle (in Polish)], *Biologia w Szkole*, 2017, pp. 43–45.

101   Kolenda et al., "Zaśmiecanie środowiska jako śmiertelne zagrożenie dla drobnej fauny" [Litter in the environment as a fatal threat to small fauna (in Polish)], *Przegląd Przyrodniczy*, vol. XXVI no 2 (2015), pp. 53–62.

102   A Ślązak, "Nabici w butelkę: śmieci wpływają na bioróżnorodność lasów" [Squeezed into a bottle: litter influences forest biodiversity (in Polish)], 03/08/2018 https://naukawpolsce.pap.pl/aktualnosci/news%2C30451%2Cnabici-w-butelke-smieci-wplywaja-na-bioroznorodnosc-lasow.html, [accessed: 09/06/2021].

103   UJ, "Sarna z plastikową butelkę na głowie błąkała się po wrocławskim lesie" [A deer with a plastic bottle on its head was wandering around a Wrocław forest (in Polish)], *Gazeta Wyborcza*, 25/04/2019, http://wroclaw. wyborcza.pl/wroclaw/7,35771,24697225,sarna-z-plastikowa-butelka-na-glowie-blaka-sie-po-wroclawskim.html, [accessed: 09/06/2021].

104   P. A. Morris, J. F. Harper, "The occurrence of small mammals in discarded bottles", *Journal of Zoology*, 1965, vol 145, issue 1, https://doi.org/10.1111/j. 1469-7998.1965.tb02010.x

105   J. J. W. Skłodowski, W. Podściański, "Zagrożenie mezofauny powodowane zaśmieceniem środowiska szlaków turystycznych Tatr", [Threats to mezofauna caused by environmental pollution on tourist trails in the Tatras (Polish)] *Parki Narodowe i Rezerwaty Przyrody*, 2004, no 23.

106   Ibidem.

107   Kolenda et al., "Survey of Discarded Bottles as an Effective Method in Detection of Small Mammal Diversity", *Polish Journal of Ecology*, 2018, no 66, pp. 57–63, https://doi.org/10.3161/15052249PJE2018.66.1.006

108   Arrizabalaga et al., "Small Mammals in Discarded Bottles: A New World Record", *Galemys*, 2016, no 28, pp. 63–65, https://doi.org/10.7325/ Galemys.2016.N4

109   Press Association, "Old Technology: NHS uses 10% of world's pagers at annual cost of £6.6 million", *Guardian*, 9/09/2017, https://theguardian.com/ society/2017/sep/09/old-technology-nhs-uses-10-of-worlds-pagers-at-annual-cost-of-66m, [accessed: 18/06/2021].

110 World Economic Forum, *A New Circular Vision for Electronics – Time for a Global Reboot*, http://www3.weforum.org/docs/WEF_A_New_Circular_Vision_for_Electronics.pdf, [accessed: 18/06/2021].

111 C. P. Baldé et al., *The Global E-Waste Monitor 2017*, https://www.itu.int/en/ITU-D/Climate-Change/Pages/Global-E-Waste-Monitor-2017.aspx, [accessed: 18/06/2021].

112 GSMA, *The Mobile Economy 2021*, https://www.gsma.com/mobileeconomy/, [accessed: 27/01/2022].

113 Statista, "Number of Smartphones sold to end users worldwide from 2007 to 2021", https://www.statista.com/statistics/263437/global-smartphone-sales-to-end-users-since-2007/, [accessed: 27/09/2021].

114 B. Merchant, "Everything That's Inside Your iPhone", *Vice*, 15/08/2017, https://www.vice.com/en_us/article/433wyq/everything-thats-inside-your-iphone, [accessed: 18/06/2021].

115 C. W. Schmidt, "Unfair Trade: e-Waste in Africa", *Environmental Health Perspectives*, 2006, issue 114(4), https://www.ncbi.nlm.nih.gov/pmc/articles/PMC1440802, [accessed: 18/06/2021].

116 P. Beaumont, "Rotten eggs: e-waste from Europe poisons Ghana's food chain", *Guardian*, 24/04/2019, https://theguardian.com/global-development/2019/apr/24/rotten-chicken-eggs-e-waste-from-europe-poisons-ghana-food-chain-agbogbloshie-accra, [accessed: 18/06/2021].

117 World Economic Forum, *A New Circular Vision for Electronics – Time for a Global Reboot*, op. cit., http://www3.weforum.org/docs/WEF_A_New_Circular_Vision_for_Electronics.pdf, [accessed: 18/06/2021].

118 Basel Action Network, *Holes in the Circular Economy: WEEE Leakage from Europe*, 2019, http://wiki.ban.org/images/f/f4/Holes_in_the_Circular_Economy-_WEEE_Leakage_from_Europe.pdf, [accessed: 18/06/2021].

119 V. Vaute, "Recycling is Not the Answer to the E-Waste Crisis", *Forbes*, 29/10/2018, https://www.forbes.com/sites/vianneyvaute/2018/10/29/recycling-is-not-the-answer-to-the-e-waste-crisis/#186f135f7381, [accessed: 02/07/2021].

120 Rreuse, *Activity Report 2017*, 2018, p11. https://www.rreuse.org/wp-content/uploads/RReuse-AR-draft4-FINAL.pdf, [accessed: 02/07/2021].

121 Ibidem, p.3.

122 "Livermore's Centennial Light Live", online webcam, http://centennialbulb.org/cam.htm, [accessed: 02/07/2021].

123 The Merzbau, a large statue, the first version of which Schwitters created in his home. A phallic block, full of nooks called grottos, in which the artist placed various items belonging to friends, but also ones vaguely associated with famous historical figures. The grottos contained bottles of urine, hairs, pictures, personal effects. The work evolved over time; Schwitters, who was deemed a degenerate artist by the Nazi regime, was forced to leave Germany and restarted his work on the Merzbau twice more.

124 H. Richter, *Dada: Art and Anti-Art,* translated by David Britt, Thames & Hudson, 1964. pp. 138–139.

125   N. Pope, "How Arman Opened the Door for Assemblage Art", *Artspace*, 21/11/2013, https://www.artspace.com/magazine/art_101/close_look/close_look_arman-51788, [accessed: 02/07/2021].

126   L. Pereira, "An Art Exhibition Mistaken for Garbage Ends up in the Trash", *Widewalls*, 28/10/2015, https://www.widewalls.ch/art-exhibition-mistaken-for-garbage/, [accessed: 02/07/2021].

127   M. Vanden Eynde, "Revolve Magazine: Artist profile", https://www.maartenvandeneynde.com/?rd_publication=153-2&lang=en, [accessed: 27/08/2021].

128   R. Nuwer, "Future Fossils: Plastic Stone", *New York Times*, 09/06/2014, https://www.nytimes.com/2014/06/10/science/earth/future-fossils-plastic-stone.html?_r=0, [accessed: 31/08/2021.]

129   K. Jazvac, "Selected solo exhibitions", https://kellyjazvac.com/CV, [accessed: 07/07/2021].

130   K. Johnson, "Swimming to Shore", *New York Times*, 15/11/2012, https://www.nytimes.com/2012/11/16/arts/design/gabriel-orozco-asterisms-at-the-guggenheim.html, [accessed: 14/07/2021].

131   S. Laville, M. Taylor, "A million bottles a minute: world's plastic binge, as dangerous as climate change", *Guardian*, https://www.theguardian.com/environment/2017/jun/28/a-million-a-minute-worlds-plastic-bottle-binge-as-dangerous-as-climate-change, [accessed: 20/07/2021].

132   Wody Polskie, "Stop suszy! Kampania społeczna" [Stop the drought! Awareness campaign (in Polish)], video, https://youtu.be/8D5f12dfbZQ, [accessed: 20/07/2021].

133   S. Kraśnicki, "Wpływ na wody podziemne i powierzchniowe projektowanej kopalni odkrywkowej węgla brunatnego eksploatującej złoże Złoczew" [The impact on ground and surface water of the planned open-caste lignite mine exploiting Złoczew deposits (in Polish)], http://eko.org.pl/odkrywki/mpe3.php, [accessed: 20/07/2021].

134   S. Kraśnicki, "Oddziaływanie projektowanej kopalni węgla kamiennego eksploatującej złoże Lublin na wody podziemne i powierzchniowe" [The impact of the planned bituminous coal mine exploiting Lublin deposits on ground and surface water (in Polish)], https://www.greenpeace.org/static/planet4-poland-stateless/2019/07/338b05c7-201906_s_krasnicki_analiza_z%C5%82o%C5%BCe_lublin.pdf, [accessed: 20/07/2021].

135   M. Grygoruk et al., "Analysis of selected possible impacts of potential E40 International Waterway development in Poland on hydrological and environmental conditions of neighbouring rivers and wetlands – the section between Polish-Belarusian border and Vistula river", 2019, http://www.ratujmyrzeki.pl/dokumenty/E40_raport_2019.pdf [accessed: 28/09/2021].

136   BBC News, "Burberry burns bags, clothes and perfume worth millions", 19/07/2018, https://www.bbc.co.uk/news/business-44885983, [accessed 15/08/2022].

137   Pesticide Action Network UK, "Pesticide Concerns in Cotton", http://www.pan-uk.org/cotton/, [accessed: 20/07/2021].

138   F. De Falco, E. Di Pace, M. Cocca, et al., "The contribution of washing processes of synthetic clothes to microplastic pollution", *Scientific Reports*, 2019, no 9, https://doi.org/10.1038/s41598-019-43023-x

139   Ellen MacArthur Foundation, *A new textiles economy: Redesigning fashion's future*, 2017, https://ellenmacarthurfoundation.org/a-new-textiles-economy, [accessed: 31/08/2021].

140   M. Fox, "Fran Lee, whose work led to pooper-scooper law, is dead at 99", *New York Times*, 22/02/2010, https://archive.nytimes.com/query.nytimes.com/gst/fullpage-940CE0DD143BF931A15751C0A9669D8B63.html, [accessed: 20/07/2021].

141   "Quand Jacques Chirac inventait la motocrotte" [When Jacques Chirac invented the 'motocrotte' (in French)], *Zigzag* https://www.pariszigzag.fr/secret/histoire-insolite-paris/quand-jacques-chirac-inventait-la-motocrotte, [accessed: 20/07/2021].

142   L. Ackland, "Don't waste your dog's poo – compost it", *Conversation*, 27/12/2018, https://theconversation.com/dont-waste-your-dogs-poo-compost-it-107603, [accessed: 20/07/2021].

143   S. Usborne, "A lesson in packaging myths: is shrink-wrap on a cucumber really mindless waste?", *Independent*, 12/11/2012, https://www.independent.co.uk/life-style/food-and-drink/features/a-lesson-in-packaging-myths-is-shrink-wrap-on-a-cucumber-really-mindless-waste-8340812.html, [accessed: 20/07/2021].

144   Glotech, "The UK's Top Food Imports and Where They Come From", 03/09/2018, https://www.glotechrepairs.co.uk/news/the-uks-top-food-imports-and-where-they-come-from/, [accessed: 20/07/2021].

145   J.-P. Schweitzer et al., "Plastic Packaging and Food Waste – new perspectives on a dual sustainability crisis", Institute for European Environmental Policy, 2018, https://ieep.eu/publications/plastic-packaging-and-food-waste-new-perspectives-on-a-dual-sustainability-crisis, [accessed: 20/07/2021].

146   J.-P. Schweitzer, C. Janssens, "Overpackaging. Briefing for the report: Unwrapped: How throwaway plastic is failing to solve Europe's food waste problem (and what we need to do instead)", in *A Study by Zero Waste Europe and Friends of the Earth Europe for the Rethink Plastic Alliance*, Institute for European Environmental Policy (IEEP), 2018, https://ieep.eu/uploads/articles/attachments/c3cd1e91-a67e-417b-8bde-d97edcf10bdd/Over%20packaging%20fact%20sheet%20-%20Unwrapped%20Packaging%20and%20Food%20Waste%20IEEP%202018.pdf, [accessed: 31/08/2021].

147   "Only 11 Years Left to Prevent Irreversible Damage from Climate Change, Speakers Warn During General Assembly High-Level Meeting", 28/03/2019, https://www.un.org/press/en/2019/ga12131.doc.htm, [accessed: 20/07/2021].

148   S. Clayton, C. M. Manning, K. Krygsman, M. Speiser, *Mental Health and our Changing Climate: Impacts, Implications, and Guidance*, Washington, D.C.:

American Psychological Association, and eco-America, 2017, https://www.apa.org/news/press/releases/2017/03/mental-health-climate.pdf, [accessed: 20/07/2021].

149   A. Grose, "How the climate emergency could lead to a mental health crisis", *Guardian*, 13/08/2019, https://www.theguardian.com/commentisfree/2019/aug/13/climate-crisis-mental-health-environmental-anguish, [accessed: 20/07/2021].

150   Special Eurobarometer, *Climate Change*, 2019, https://ec.europa.eu/clima/sites/clima/files/support/docs/pl_climate_2019_en.pdf, [accessed: 20/07/2021].

151   V. Knight, "'Climate Grief': Fears About the Planet's Future Weigh On Americans' Mental Health", https://khn.org/news/climate-grief-fears-about-the-planets-future-weigh-on-americans-mental-health/, [accessed: 20/07/2021].

152   B. Johnson, *Zero Waste Home: The Ultimate Guide to Simplifying Your Life by Reducing Your Waste*, New York, 2013.

153   E. Thelen, "'Eco-shaming' is on the rise, but does it work?", World Economic Forum, 18/07/2019, https://www.weforum.org/agenda/2019/07/eco-shaming-is-rising-but-does-it-work/, [accessed: 20/07/2021].

154   A. Konieczna, A. Rutkowska, D. Rachón, "Health Risks of Exposure to Bisphenol A (BPA)", *Rocznik Państwowego Zakładu Higieny*, 2015, 66 (11), pp.5-11, https://www.ncbi.nlm.nih.gov/pubmed/25813067, [accessed: 22/07/2021].

155   Pt, "Polska droga do gospodarki o obiegu zamkniętym: Opis sytuacji i rekomendacje" [The Polish Route to a Circular Economy: the current situation and recommendations (in Polish)], 21/04/2017, http://www.portalsamorzadowy.pl/pliki-download/97853.html, [accessed: 28/09/2021].

156   F. Falchi et al., "The New World Atlas of Artificial Night Sky Brightness", *Science Advances*, 2016, vol. 2, no 6, https://advances.sciencemag.org/content/2/6/e1600377, [accessed: 22/07/2021].

157   "Confused Sea Turtles go to Italian Restaurant", *Telegraph*, 19/08/2008 https://www.telegraph.co.uk/news/earth/earthnews/3349881/Confused-sea-turtles-go-to-Italian-restaurant.html, [accessed: 22/07/2021].

158   T. Le Tallec, "What is the ecological impact of light pollution?", *Encyclopedia of the Environment*, 05/02/2019, https://www.encyclopedie-environnement.org/en/life/what-is-the-ecological-impact-of-light-pollution, [accessed: 22/07/2021].

159   M. Chauvot, "Coronavirus: la collecte des déchets s'adapte au confinement" [Coronavirus: waste recycling adapts to lockdown (in French)], *Les Echos*, 20/03/2020, https://www.lesechos.fr/industrie-services/energie-environnement/coronavirus-la-collecte-des-dechets-sadapte-au-confinement-1187344, [accessed: 22/07/2021].

160   M. Victory, M. Tudball, "European Plastics Recycling Markets React to

Coronavirus", *Recycling Today*, 24/03/2020, https://www.recyclingtoday.com/article/icis-assement-of-conoravirus-impacts-european-plastics-recycling/, [accessed: 22/07/2021].

161 R. Smithers, "Coronavirus: UK faces cardboard shortage due to crisis", *Guardian*, 31/03/2020, https://www.theguardian.com/environment/2020/mar/31/uk-faces-cardboard-shortage-due-to-coronavirus-crisis, [accessed: 22/07/2021].

162 RECOVER Coalition Letter, 16/04/2020, https://www.documentcloud.org/documents/6877535-RECOVER-Coalition-Letter.html, [accessed: 22/07/2021].

163 Reuters, "Discarded coronavirus masks clutter Hong Kong's beaches, trails", 12/03/2020, https://news.trust.org/item/20200312052301-62ydm/, [accessed: 22/07/2021].

164 S. McCarthy, "Coronavirus can stay on face masks for up to a week, study finds", *South China Morning Post*, 06/04/2020 https://www.scmp.com/news/china/science/article/3078511/coronavirus-can-remain-face-masks-week-study-finds, [accessed: 22/07/2021].

165 N. van Doremalen et al., "Aerosol and Surface Stability of SARS-CoV-2 as Compared with SARS-CoV-1", *New England Journal of Medicine*, 2020, vol. 382:1564-1567, https://www.nejm.org/doi/full/10.1056/NEJMc2004973, [accessed: 22/07/2021].

166 European Commission, "Waste management in the context of the coronavirus crisis", 14/04/2020, http://ec.europa.eu/info/sites/info/files/waste_management_guidance_dg-env.pdf, [accessed: 22/07/2021].

167 M. Bednarek, "W Zawierciu afera. Pod domami ustawiono kosze na śmieci z napisem ‚COVID-19" [Scandal in Zawiercie: bins labelled ‘COVID-19' installed outside homes (in Polish)], *Gazeta Wyborcza*, 29/04/2020, https://katowice.wyborcza.pl/katowice/7,35063,25908557,pod-domami-ustawili-kosze-na-smieci-z-napisem-covid-19-wybuchla.html, [accessed: 22/07/2021].

168 Kk/eŁKa/pm/ks, "Sugerował oznaczenia na domach osób zakażonych koronawirusem. Starosta: byłem tylko przekaźnikeim" [He suggested marking the homes of people infected with coronavirus. The prefect: I was only the messenger (in Polish)], *TVN24*, video, 18/04/2020, https://tvn24.pl/bialystok/wysokie-mazowieckie-starosta-bogdan-zielinski-proponuje-oznaczanie-domow-osob-zakazonych-koronawirusem-4559094, [accessed: 22/07/2021].

169 European Plastics Converters, "Open Letter", 08/04/2020, https://fd0ea2e2-fecf-4f82-8b1b-9e5e1ebec6a0.filesusr.com/ugd/2eb778_9d8ec284e39b4c7d84e774f0da14f2e8.pdf, [accessed: 22/07/2021].

170 Plastic Industry Association, 18/03/2020, https://www.politico.com/states/f/?id=00000171-0d87-d270-a773-6fdfcc4d0000, [accessed: 22/07/2021].

171 J. Calma, "Plastic Bags are making a comeback because of COVID-19", *Verge*, 02/04/2020, https://www.theverge.com/2020/4/2/21204094/plastic-bag-ban-reusable-grocery-coronavirus-covid-19, [accessed: 22/07/2021].

STANISŁAW ŁUBIEŃSKI first began observing birds in childhood through Soviet binoculars. He is a regular contributor to newspapers and magazines, and his first book to be translated into English, *The Birds They Sang*, won the readers' vote for the Nike Literary award, Poland's most prestigious literary prize.

ZOSIA KRASODOMSKA-JONES translates primarily from Polish into English. Previous translations include *The Book of Dirt* by Piotr Socha and Monika Utnik-Strugała, and *Mud Sweeter than Honey* by Margo Rejmer, a co-translation with Antonia Lloyd-Jones.